Let's Talk Safety

52 Talks on Common Utility Safety Practices for Water Professionals

2025

American Water Works Association

Let's Talk Safety: 52 Talks on Common Utility Safety Practices for Water Professionals

Printed in the United States of America.

Senior Editorial Manager – Books: Suzanne Snyder
Senior Acquisitions Editor: Geoffrey S. Shideler
Production: Gillian Wink
Cover Illustration: Lerbank-bbk22/Shutterstock.com

Every attempt has been made to ensure that the contents of this book are up-to-date and follow the latest safety guidelines and regulations. However, such guidelines and regulations are subject to frequent changes and variation by state, province, or region. The authors, contributors, editors, and publisher do not assume responsibility for the validity of the content or any consequences of its use. In no event will AWWA be liable for direct, indirect, special, incidental, or consequential damages arising out of the use of information presented in this book. In particular, AWWA will not be responsible for any costs, including, but not limited to, those incurred as a result of lost revenue. In no event shall AWWA's liability exceed the amount paid for the purchase of this book.

ISBN   978-1-64717-188-9
ISBN, electronic   978-1-61300-715-0

American Water Works Association

6666 West Quincy Ave
Denver, CO 80235-3098
303.794.7711

www.awwa.org

Contents

How to Use *Let's Talk Safety*

The American Water Works Association is pleased to provide you with this edition of *Let's Talk Safety*. AWWA Health, Safety, and Environment Committee members have reviewed and suggested updates to the discussion topics to ensure they continue to be current, pertinent, and beneficial to you and your employees.

Website addresses at the end of nearly every article allow you and your staff to conduct deeper research into particular safety topics.

Refer to the end of this book for a selection of standards and manuals, as well as information on AWWA's Safety First Video Series available through streaming. These materials can be used in conjunction with this publication to create a holistic approach to company-wide safety training.

We are all seeking an injury-free work environment. Our universal goal is to have every employee, every day, return home to his or her family uninjured. We believe that the first step to not being injured is knowing that you can be injured. *Let's Talk Safety* is designed to help you build awareness of potential work hazards and provide safety practices that help mitigate those hazards. Talking to your employees about their safety and listening to their safety concerns and experiences are the foundation of building an effective safety culture. This book will help you initiate important safety dialogues by providing common starting points for discussion. You may also want to consider placing these articles in your employee safety publications.

Here's how to make your safety meetings more engaging and effective:

- Cover only one safety topic in a meeting. Employees can easily lose focus when too many topics are discussed.

- Ensure the discussion topic is pertinent to the participants. *Let's Talk Safety* covers topics in a generic manner, and a particular talk may not apply to every workplace and every work situation. Be creative and use a topic presented here and relate it to your work group's particular safety issue or concern.

- Provide examples. Citing real accidents that are in the news or something that occurred at another facility (or your own) can provide a reality check for the topic and take it out of the realm of "It'll never happen to me."

- Involve the employees in the meeting. You may want to appoint a different employee each week to lead the discussion. Ask questions and ask for personal examples of near misses and hazardous situations.

- Don't let a safety meeting become a complaint session—especially if it's not about safety! Acknowledge the complaint, and let the workers know it will be addressed afterward. Keep the focus on the safety topic at hand.

- Chalkboards, charts, videos, and other interactive materials will help keep the topics engaging. Change up the meetings occasionally by bringing in the tools or personal protective equipment being discussed. When talking about large equipment, hold the meeting in the yard and use the specific equipment as the backdrop.

- Occasionally invite guest speakers who are experts in a particular subject.

- Conduct your meetings early in the week so employees have a chance to practice what they hear.

- Avoid embarrassing a particular employee by pointing out that person as an example of what not to do. Speak in generalities if possible.

- Distribute copies of the *Let's Talk Safety* briefing each week.

The safety awareness information presented in this book is designed to help your utility workers develop a greater awareness of potential job hazards and help them make informed, judicious decisions. The information contained in *Let's Talk Safety* provides only general safety awareness guidelines related to the many aspects of working in the water utility industry. This compendium is not comprehensive and does not cover every potential aspect of a safety issue a typical water utility worker may encounter.

The safety articles are not intended, nor should they be considered, as a substitute for more comprehensive and formal safety training courses and certification programs that provide greater detail and explanation.

For employees to do their jobs effectively and safely, they must be responsible for learning and understanding the safety rules and regulations that apply to their particular occupation. Health and safety regulations and requirements mandated by federal, state, and local governments, as well as by your company's established policies and regulations, need to be consulted before any work begins.

This book is updated as necessary to remain current with OSHA and other guidelines and provide safety recommendations accordingly.

Acronyms and Abbreviations

AED automated external defibrillator

AHA American Heart Association

AWWA American Water Works Association

CDC Centers for Disease Control and Prevention

CO carbon monoxide

CPR cardiopulmonary resuscitation

dB decibels

EMS emergency medical service

GHS Globally Harmonized System of Classification and Labeling of Chemicals

HCS Hazard Communication Standard

HIV human immunodeficiency virus

JHA job hazard analysis

LEL lower explosive limit

LFL lower flammable limit

MSDS Material Safety Data Sheet (now SDS)

NHTSA National Highway Traffic Safety Administration

NFPA National Fire Protection Association

NIOSH National Institute for Occupational Safety and Health (a division of CDC)

NRR noise reduction rating

OSHA Occupational Safety and Health Administration

PEL permissible exposure limit

POC point of contact

PPE personal protective equipment

ROPS rollover protective structure

SCA sudden cardiac arrest

SDS Safety Data Sheet

TLV threshold limit value

USEPA US Environmental Protection Agency

WARN Water/Wastewater Agency Response Network

How to Conduct a Safety Tailboard

The first step to being injury-free is knowing that you can be injured, regardless of how safe you think you may be. On-the-job safety is typically not your only work goal, but it must be the highest work priority—for both you and your coworkers. Employees are often forced to reconcile between competing goals: timeliness versus safety. There really isn't a choice. You must always choose your safety and the safety of others ahead of everything else. If you see an unsafe work situation, you owe it to yourself and to your coworkers to immediately stop the work until the situation is made safe!

Most jobsite injuries happen with new workers who don't know the safety rules and with older workers who become complacent about established safe work practices. In essence, these veteran workers have learned over the years to take shortcuts in established safety procedures. These shortcuts eventually become the working norm and get passed through the workforce.

Within the utility industry, a standard work practice for field crews is to conduct a safety tailboard session before the work begins. While this chapter focuses on field crews, the basic principles apply to office projects as well.

A safety tailboard session is about good communication. It helps everyone involved in the project to fully understand the processes and procedures, and it can effectively reduce injuries.

Plan the Work

Basics of an effective tailboard include the following.

- It is conducted by the employee in charge.

- Simple and plain language is used so that everyone knows and understands exactly what is being said. Encourage questions.

- Hold the meeting just before work begins and again if any significant changes to the jobsite occur.

- Review all applicable safety rules regarding your company's procedures and the required personal protective equipment (PPE).

- Make a safety plan and an emergency plan—even if working alone!

- Analyze the job's processes and procedures and discuss what safety and rescue issues could come into play if there is an accident.

Work the Plan

- Discuss the potential hazards and special precautions that the job requires and that the jobsite might provide.

- Discuss the job's processes, procedures, and tasks to be performed and in what order they will be performed. Always include a review of all appropriate safety procedures and considerations.

- Discuss everyone's assignments. Make sure all know their jobs and the jobs of their coworkers.

- Establish a worker buddy system where coworkers are assigned to watch out for each other when in a remote location.

- Ensure that those with new job assignments or new tools or equipment are properly and completely trained in the safety processes, procedures, and tool or equipment operation.

- Conduct inspections whenever new substances, processes, procedures, tools, or equipment are introduced and may present a safety issue.

- Discuss the tools and PPE needed to complete the jobs safely and without incident. Inspect the tools for proper and safe operation. Ensure all PPE is up to standard and safe to use.

- Report hazards and unsafe equipment to the supervisor before work begins.

- Discuss unusual and nonroutine situations.

- Discuss emergency procedures. Determine ahead of time who is in charge in an emergency situation and who is the backup.

- Know where all emergency resources are located: emergency plan, fire extinguisher, first-aid and burn kits, and communication devices such as a phone, cell phone, or radio.

- Before an accident occurs, discuss how to direct emergency services to your location.

In addition to the topics presented in this edition of *Let's Talk Safety*, numerous commercial and educational websites offer safety meeting discussion materials and topics.

Climbing Elevated Tanks:
The Height of Safety

The dangers of climbing elevated water storage structures should never be underestimated. Utility staff often must climb structures higher than 12 ft when climbing towers to check paint, look for rust or other damage, and inspect hatches, locks, and beacon lights. Without protection, the workers risk falling several stories. Even if a worker is correctly anchored, a fall in a safety harness can cause whiplash and a loss of circulation known as suspension trauma. Injury or sudden illness could also incapacitate an employee while he or she is working on a tower, requiring an emergency evacuation.

Source: Magda Wygralak/Shutterstock.com

A qualified fall-protection trainer can teach staff proper climbing techniques and how to use personal fall-arrest equipment and harnesses, as well as how to use personal fall-arrest equipment and harnesses, as well as how to correctly handle a fall.

Some water structures have fall-arrest systems on their ladders: a climber connects the locking sleeve directly to the D-ring on the harness the climber wears for protection. On older structures, however, climbers manually connect snap-hook lanyards onto the ladder's side rails—not its rungs—and maintain three points of contact (both hands and one alternating foot) while moving.

The transitions from ladder onto overhead catwalk or from ladder through a hatch (and vice versa) are the most dangerous parts of any climb. Climbers should always attach a fall-arrest lanyard to an approved anchor point before making a transition or while working topside around an open hatch or near the edge.

The physical exertion involved in utility-tower climbing should not be underestimated. For the average person in reasonably good condition, it can be a full-body workout—especially if one is carrying an extra load, such as electricians' tools attached to 15 or so pounds of harness and other PPE.

Climbers should always use the buddy system. Someone, even a non-climber on the ground, should be onsite to phone 911 immediately if a climber gets into trouble and cannot get down.

Standard operating procedures for using any fall protection system should include the following safety guidelines.

- Only personnel who are authorized to climb and have completed an authorized person training with practical exercises are allowed to use fall-protection systems.

- Climbers must inspect and then don proper fall-arrest equipment, including a full-body harness, double lanyards with one-hand operation, and an ascender/descender (sleeve) device if the structure is equipped with a fall-arrest system in good working condition.

- An appropriate ANSI Z89-1 climbing helmet or hard hat must be worn at working height and on the ground.

- Climbers should never ascend a structure while onsite alone. At a minimum, an employee with a cellphone or radio should be stationed on the ground, with the climber in visual and/or shouting range. Otherwise, climbers should employ a buddy system of two or more trained personnel at altitude.

Additionally, Rope and Rescue School, whose motto is Knowledge = Safety, provides these tips to tower climbers:

- Don't be cocky or a show-off or have a competitive attitude when working at heights.

- If you are tired, take a rest. Fatigued muscles don't respond as quickly.

- Do not work above people and do not let people work above you.

- Warm up. You'll feel stronger and lighter, protect yourself from injuries, and improve your aerobic threshold and general endurance.

- Empty your pockets of objects that can potentially turn into projectiles.

- Start hydrated and stay hydrated.

- Stay 100 percent tied in while climbing, descending, working in position, and maneuvering around the tower.

- If an emergency arises, the ground-safety staffer or fellow climber is responsible for phoning 911. The emergency caller must specify the address of the emergency, describe the nature of the problem, and identify the urgent need for "high-angle rescue and EMS."

- If a climber gets into trouble and is incapacitated, the second person must not leave the structure until the stricken climber is down. The ground-safety staffer or fellow climber should provide rescue personnel with an approximate duration of time since the climber fell to help assess the medical effects of restricted blood circulation in the victim's limbs from hanging in a full-body harness.

- If a climber slips and falls, engaging the fall-arrest system, his or her body harness (and lanyard, too, if used) has been "shock loaded." After the climber returns to the ground, the harness can no longer be worn and must be taken out of service, as specified by the PPE manufacturer.

For more information, visit the OSHA website on fall protection: **www.osha.gov/ fall-protection**

Understanding Safety Data Sheets

When you work with chemicals in the workplace, you need to have detailed information on the particular hazardous chemical(s) you are working with. You also need guidelines for appropriate protective measures, handling and storage, and preventing or mitigating spills, fires, or injuries. Container labels don't always tell you everything you need to know about hazardous materials.

The OSHA Hazard Communication Standard (HCS) requires chemical manufacturers, distributors, and importers to provide Safety Data Sheets (SDS; formerly known as Material Safety Data Sheets, or MSDS) to inform users about chemical product hazards. Employers must ensure that SDS are readily accessible to employees. As of June 1, 2015, new SDS are required to be in a uniform format (Globally Harmonized System of Classification and Labeling of Chemicals, or GHS), with the section numbers, headings, and specifics included below.

1. Identification: Product identifier; manufacturer or distributor name, address, phone number; emergency phone number; recommended use; restrictions on use

2. Hazard(s) identification: All hazards regarding the chemical; required label elements

3. Composition/information on ingredients: Information on chemical ingredients; trade secret claims

4. First-aid measures: Important symptoms/effects, both acute and delayed; required treatment

5. Fire-fighting measures: Suitable extinguishing techniques, equipment; chemical hazards from fire

6. Accidental release measures: Emergency procedures; protective equipment; proper methods of containment and cleanup

7. Handling and storage: Precautions for safe handling and storage, including incompatibilities

8. Exposure controls/personal protection: OSHA's permissible exposure limits (PELs), threshold limit values (TLVs), appropriate engineering controls, and PPE

9. Physical and chemical properties: The chemical's characteristics

10. Stability and reactivity: Chemical stability and possibility of hazardous reactions

11. Toxicological information: Routes of exposure, related symptoms, acute and chronic effects, numerical measures of toxicity

12. Ecological information*: Information provided here helps environmental professionals in the event of a release

13. Disposal considerations*: Provided here is information about the chemical classification for waste-disposal laws

14. Transport information*

15. Regulatory information*: This section contains information about the regulatory status of the material for OSHA and other federal agencies

16. Other information: Date of preparation or last revision

*Other (non-OSHA) agencies regulate sections 12–15.

For more information, visit **www.osha.gov/hazcom**

Powerful Protection From PPE

You wouldn't think of wearing a parka to waterski or a tuxedo to install a sump pump. If you think these are examples of extreme fashion gaffes, think again. A far more serious misstep is tackling a job without wearing the right personal protective equipment (PPE). PPE is designed to protect the eyes, face, head, respiratory tract, and body extremities from potentially hazardous conditions. It includes such items as safety glasses, goggles, face shields, hard hats, respirator, cut or puncture resistant gloves, protective clothing, welding aprons, and safety shoes, to list just a few. The workplace (or jobsite) must be assessed to determine if hazards are, or may be, present that will require PPE use. The right PPE must be selected and employees fitted and trained in its proper use. Let's briefly review some of the most common PPE.

Source: Rawf8/Shutterstock.com

Since you never know when you might be diverted to a hazardous task, some facilities specify "always used" PPE (e.g., safety glasses, protective footwear, gloves, and hard hat). Review your workplace (or jobsite) job hazard assessments or guides that must be based on assigned tasks and hazards that are, or may be, present that will require PPE use. Know your job's PPE requirements! The right PPE must be selected and employees fitted and trained in its proper use. Let's briefly review some of the most common PPE.

Eye and Face Protection

Eye and face protection is necessary when there is potential exposure to flying particles and dust (wood, glass, metal); molten metal (welding spatter); potentially injurious light radiation (welding glare); or chemicals in any form—liquid, vapor, or gaseous. Eye and face PPE can include safety glasses with side shields, chemical goggles, or a full-face respirator. All devices must comply with federally referenced consensus standards.

Remember, not all eye or face protection will protect you from all hazards! Protective glasses with side shields are fine for particulates but provide no protection from hazardous chemical vapors. As more treatment facilities are using Ultraviolet Light disinfectant UV C rated face shield and safety glasses should be used. The PPE must fit the hazard. **www.osha.gov/eye-face-protection**

Respiratory Protection

The first step in controlling potentially hazardous dusts, mists, fumes, smoke, or gases in the workplace is the installation of engineering controls such as mechanical ventilation systems. If such measures are infeasible (e.g., at a field work-site), then respiratory PPE must be used. A respirator may also be specified as a precaution against accidental release (e.g., changing a chlorine container).

The two basic categories of respirators are air-purifying and atmosphere-supplying. The first device simply filters the ambient air using an air-purifying filter, cartridge, or canister. The second type actually provides breathing air to the user from an independent source. The category of respirator; style (half, full-face, or Powered Air Purifying Respirator), and type of filter or canister all must be carefully selected. Employees must be properly fitted for PPE and trained in when and how to use it. A medical evaluation of a person's ability to effectively wear and use a respirator must also be conducted: **www.osha.gov/respiratory-protection**

Head Protection

When working in an area where the potential exists for head injuries resulting from falling objects or impact hazards (e.g., crane in use, overhead levels not isolated by walls and floors), employees must wear head protection, which usually comes in the form of hard hats. Again, as with other forms of PPE, hard hats must be manufactured to federal standards and worn properly to afford proper head protection. **www.osha.gov/personal-protective-equipment**

Foot Protection

Where there are potential hazards to the feet from falling or rolling materials, sharp objects that can pierce the sole, or electrical shock, employees must wear appropriate protective footwear. This footwear commonly takes the form of safety shoes, often equipped with steel toes and shanks and heavy-duty, slip resistant soles: **www.osha. gov/personal-protective-equipment**

Hearing Protection

Hearing is a precious gift. Recurring exposure to elevated noise levels can seriously damage your hearing. Workplaces with average noise exposures above 85 dbA averaged over 8 working hours must have a Hearing Conservation Program. If noise levels are too high, employees must be supplied with and must use hearing protection. Hearing protection provided can be disposable earplugs or reusable ear-muffs. In extreme cases of noise, both may be required. The protection needed depends on the nature of the hazard and duration of exposure on the job: **www.osha.gov/noise**

Other PPE

Source: Aleksandar Malivuk/Shutterstock.com

Other PPE can take the form of gloves, welding aprons, chemical protective suits, coveralls, and back support braces. All are designed to protect a very important person—you—from potential hazards you might encounter on the job.

But remember, no PPE will protect your vision, your lungs, your head, or any other part of your body unless you wear it and wear it correctly. Be fashionable—be safe!

For additional information, go to the OSHA website on PPE: **www.osha.gov/ personal-protective-equipment**.

Safety Tips for Employees Working Remotely or Alone

Reductions in manpower and increases in workload have increased the number of field employees who are working alone. While lone work may not automatically decrease a worker's safety performance, there is no doubt that working alone increases a worker's vulnerability to a variety of safety issues. This vulnerability applies not only to those who regularly work outdoors but also to any employees whose work frequently takes them out into the community. All these workers may encounter threats to their safety.

The following four steps can help you reduce the safety vulnerability of remote workers. These tips apply to all employees and their managers who must work alone or with others in remote locations where normal means of communication are unreliable or nonexistent.

Routine Communications Protocol

- Designate a key point of contact (POC) who is not a part of the remote team.
- Know who is working remotely and how long the work should take.
- Set regular check-in times for the entire work period. At each check-in, the POC should record the time and the information given by the remote worker.
- Evaluate lighting conditions; are they sufficient to ensure worker safety?
- POCs should relay any anticipated changes in weather.

Emergency Communications Protocol

- If a check-in time is missed, the POC should try, for 30 minutes, to reestablish communications.

If that fails, the POC must do the following.

- Assemble a search team and place the team on standby.
- Contact medical personnel, informing them that an emergency response may be needed. If an event includes an injury, after ensuring that medical attention

has been provided, the responsible supervisor shall ensure that the appropriate incident/injury reporting process is initiated.

Evaluate the Potential Hazards

Before entering a remote work location, all team members should identify and discuss the following potential safety issues:

- planning for weather conditions—both forecast and unexpected;
- facing potential emergencies such as flooding, electrical contact, running out of fuel in cold climates, and so on;
- handling serious injuries or illnesses that might occur far from medical facilities;
- guarding against animal attacks, snakes, and insects;
- making contact with emergency agencies;
- having the appropriate PPE; and
- having the tools required to complete the job safely.

Team members should also assess the risks and review work-related documentation such as a job hazard analysis to ensure all mitigation and control measures have been addressed.

Conduct a Safety Tailboard

- Discuss potential hazards and special precautions the work requires.
- Discuss the job's processes, procedures, and tasks and the order in which they will be performed.
- Review appropriate safety procedures and PPE considerations. Inspect tools and ensure all PPE meets safety standards.
- Discuss assignments. All must know their jobs and the jobs of their coworkers.
- Establish a buddy system where coworkers watch out for each other.
- Ensure that those with new job assignments, new tools, or new equipment are properly and completely trained in safety processes, procedures, and tool/equipment operation.
- Everyone should regroup and discuss potential safety issues when new substances, processes, procedures, tools, or equipment are introduced to the worksite.
- Report hazards and unsafe equipment to the supervisor before work begins.
- Discuss unusual and non-routine situations.

- Discuss emergency procedures. Determine ahead of time who's in charge in an emergency situation and who is the backup.

- Know where all emergency resources are located: emergency plan, fire extinguisher, first-aid and burn kits, and communication devices.

For additional information and ideas, see Service NL (Newfoundland Labrador) Working Alone Safely Guidelines for Employers and Employees: **www.servicenl.gov. nl.ca/ohs/safety_info/si_working_alone.html**

CPR and AEDs Can Save Lives

Sudden cardiac arrest (SCA) is the sudden, unexpected loss of heart function, breathing, and consciousness. SCA occurs when the heart's electrical function—its ventricular fibrillation—is interrupted and stops the heart from pumping. SCA can also occur with a heart attack, which occurs when blood flows to a portion of the heart is blocked. Either way, without medical attention, the victim will die.

Of the nearly 300,000 people in the United States who suffer an out-of-hospital SCA, 92% die, according to the Centers for Disease Control and Prevention. What survivors have in common are early intervention with cardiopulmonary resuscitation (CPR) and an automatic external defibrillator (AED), followed by rapid delivery of appropriate care—usually a trip to the emergency room. **www.osha.gov/aed**

SCA can be caused by

- heart attack and other cardiac conditions
- electrocution
- asphyxiation (loss of consciousness and death caused by inadequate oxygen in the work environment, such as in a confined space); and
- trauma, drowning, overdose, primary respiratory arrests, anaphylactic shock, and other noncardiac conditions.

Many victims have no prior history of heart disease and are stricken without warning. When someone suddenly loses consciousness, think CCCC: Clear, Check, Call, and Compress.

- *Clear* the area of other safety hazards. Make sure that the victim and you are safe from further harm.
- *Check* the victim for responsiveness. Has he stopped breathing or is he gasping irregularly for air? Does he respond at all to a hard slap on the shoulder blades?

- *Call* for help. If someone else is around, tell her to call 911 and find the nearest AED, if one is available. AEDs provide an electric shock that can restore normal rhythm to a heart in ventricular fibrillation.

- *Compress* the chest hard and fast. Push straight down on the lower sternum, using one hand on top of the other at the rate of 100 times a minute.

Compressions are the most important part of CPR. New American Heart Association (AHA) guidelines no longer require the rescuer to provide life-saving breaths to the victim because compressions, done properly, will keep the blood circulating throughout the victim's body. There is enough oxygen in the blood of the victim to keep the heart, brain, and organs alive if it is circulated through chest compressions, and time spent assessing breathing is better spent compressing. This is known as hands-only CPR. For more information, go to **http://cpr.heart.org/en**

However, if you can provide breaths, or medical help isn't immediately available, then

1. Open the airway with a gentle head tilt and chin lift,

2. Pinch the victim's nose closed, and

3. Take a normal breath, cover the victim's mouth with yours to create an airtight seal, and give two 1-second breaths as you watch for the chest to rise.

Staff trained in CPR and the use of an AED can save precious treatment time and improve survival odds because they provide aid before emergency medical service (EMS) personnel arrive.

Basic CPR can be learned in less than a day of training, and many businesses will either sponsor their staff to attend CPR classes or bring a professional in for staff training. A person trained in CPR can assess whether a victim needs to be treated with chest compressions and airway breaths and then appropriately conduct the following procedures.

Here's a step-by-step guide for the latest CPR:

1. Check the scene for safety, form an initial impression, and use personal protective equipment.

2. If the person appears unresponsive, check for responsiveness, breathing, life-threatening bleeding, or other life-threatening conditions using shout-tap-shout.

3. If the person does not respond and is not breathing or only gasping, call 911 and get equipment, or tell someone to do so.

4. Kneel beside the person. Place the person on their back on a firm, flat surface.

5. Give 30 chest compressions.

   - Hand position: Two hands centered on the chest
   - Body position: Shoulders directly over hands; elbows locked
   - Depth: At least 2 inches
   - Rate: 100 to 120 per minute
   - Allow chest to return to normal position after each compression

6. Give two breaths.

   - Open the airway to a past-neutral position using the head-tilt/chin-lift technique.
   - Pinch the nose shut, take a normal breath, and make a complete seal over the person's mouth with your mouth.
   - Ensure each breath lasts about 1 second and makes the chest rise; allow air to exit before giving the next breath.

Note: If the first breath does not cause the chest to rise, re-tilt the head and ensure a proper seal before giving the second breath. If the second breath does not make the chest rise, an object may be blocking the airway.

7. Continue giving sets of 30 chest compressions and 2 breaths. Use an AED as soon as one is available! Minimize interruptions to chest compressions to less than 10 seconds. **www.redcross.org/take-a-class/cpr/performing-cpr/cpr-steps**

8. When the AED arrives, turn it on and follow the audio prompts.

Don't worry about pushing too hard. In fact, some rib bones will probably crack or break if you are correctly compressing the heart. Ribs are repairable, but when the heart stops, the absence of oxygenated blood can cause permanent brain damage within minutes. Death will occur within 8–10 minutes. For every minute that treatment is delayed, the survival rate drops 10%. The earlier CPR is initiated, the greater the chances of survival.

If help is provided within four minutes, the chances of survival are doubled. These few minutes can be the difference between life and death.

Note: The AHA still recommends breaths with compressions for infants and children and victims of drowning or drug overdose, or for people who collapse due to breathing problems.

For more information, go to the American Heart Association website: **www.heart.org**, or the American Red Cross website: **www.redcross.org**

Songs to Save a Heart

Each of these songs has a rhythm of about 100 beats a minute.

"Stayin' Alive" by The Bee Gees	"Sweet Home Alabama" by Lynyrd Skynyrd
"I Will Survive" by Gloria Gaynor	"Mmmbop" by Hanson
"Ob-La-Di, Ob-La-Da (Life Goes On)" by The Beatles	"Girls Just Wanna Have Fun" by Cyndi Lauper
"Kickstart My Heart" by Motley Crue	"Another Brick in the Wall" by Pink Floyd
"Heartbreaker" by Mariah Carey	"Gives You Hell" by The All American Rejects
"Achy Breaky Heart" by Billy Ray Cyrus	"Crazy" by Gnarls Barkley
"Body Movin'" by the Beastie Boys	"People Are Strange" by The Doors
"Rock Your Body" by Justin Timberlake	"What's Going On" by Marvin Gaye
"Hard To Handle" by The Black Crowes	"Dancing Queen" by ABBA
"Breathless" by Maroon 5	"Another One Bites The Dust" by Queen
"Do You Really Want To Hurt Me?" by Boy George	"Jingle Bells"
"Cecilia" by Simon and Garfunkel	"Nellie the Elephant"

Energized Electric Equipment Can Be Deadly

Utility workers often encounter situations in which they are required to work with energized electric tools or equipment. The most important thing to remember in these situations is to always consider the electric circuits, apparatus, and your tools to be energized and deadly. On average, a construction worker is electrocuted and killed once a day somewhere in the United States. And more than 3,000 field workers are severely burned or injured every year by electrical mishaps on the jobsite.

Electricity can hurt, burn, and kill you—even at low voltages. Always keep in mind that electricity travels at the speed of light and that it is trying to find the path of least resistance to get to ground. Your body is made up mostly of water and therefore is an excellent conductor of electricity. The effects of an electrical current passing through the body range from a mild tingling sensation to severe pain, muscular contractions, to even death. As the current passes through a body, it will burn from the inside out at about 6,000°F (3,315°C).

Beware of Overhead Power Lines

Before you begin work, survey the jobsite to find overhead power lines, poles, and guy wires. Look for lines that may be hidden by trees or buildings. Conditions change, so check daily.

- Point out power lines at the daily work briefings.

- Assume all overhead lines are energized and potentially dangerous, including service drops that run from utility poles to buildings.

- Remember the 10-ft rule: Keep vehicles, equipment, tools, scaffolding, and people at least 10 ft away from overhead power lines.

- If you must work closer than 10 ft, contact your local electric utility in advance to make safety arrangements.

- Higher-voltage power lines require greater clearance. Contact your local electric utility for specific clearances.

- Clearly mark boundaries to keep workers and equipment a safe distance from overhead lines.

- Use a spotter. Equipment operators need a designated spotter who can help keep you clear of power lines and other safety hazards.

- Call 811 before you dig.

Call your local dig alert service at 811 at least two working days before digging. If you don't call and you hit an underground line, you could be hurt or killed. You may also be liable for costly damages.

Source: Ivan Bruno de M/Shutterstock.com

Avoiding Electrical Accidents and Shock

The easiest way to avoid electrical accidents is simply to avoid contact with energized components. Always assume that an electrical circuit is energized and dangerous until you are certain that it is not. Before working on a circuit, use a voltage meter to determine if the circuit is energized.

Before you work on electrical equipment, turn off the power to it. Use your standard lockout/tagout procedures before you begin working anywhere near the energized equipment.

To be safe, all electrical equipment and apparatuses must be double-insulated or grounded. If possible, avoid the use of extension cords. When extension devices (an enclosure with multiple sockets) must be temporarily used, the wire gauge of the device must be equal to or larger than the cord on the item being operated. Never attach extension devices to building surfaces using staples, nails, or similar means.

Extension devices equipped with surge protectors can be permanently used with equipment that contains microprocessors, such as computers, but surge protectors should not be used in areas subject to moisture, physical or chemical damage, or flammable vapors.

Follow these simple safeguards to avoid electric shock:

- Check your work area for water or wet surfaces near energized circuits. Water acts as a conductor and increases the potential for electrical shock.

- Check for metal pipes and posts that could become the path to ground if they are touched.

- Do not wear rings, watches, or other metal jewelry when performing work on or near electrical circuits. They are excellent conductors of electricity.

- Leather gloves will not protect you from electrical shock. They are cowhide, typically, and have inherent moisture in them.

- Never use metal ladders or uninsulated metal tools on or near energized circuits.

- Make it a daily habit to examine your electrical tools and equipment for signs of damage or deterioration. Do not use them if the electrical wires are damaged or if they are not insulated or grounded. Defective cords and plugs should be discarded immediately and replaced.

For more information, go to OSHA's Safety and Health Topics page on electrical equipment: **www.osha.gov/SLTC/electrical**, or visit the Electrical Safety Foundation International website: **www.esfi.org**. Reach out to your local electric utility for specific safety information.

Trenching: Don't Dig Into Trouble!

If you're involved with water utility maintenance or construction, sooner or later you're going to be involved in trenching operations. And despite all the classic slapstick movie routines you may have seen through the years, safely excavating and working in an open trench are serious business.

Not all holes in the ground are trenches. A trench is defined as a narrow excavation made below the surface of the ground. In general, the depth is greater than the width, as measured at the bottom. Trenches 5 ft deep or greater require a protective system unless the excavation is made entirely in stable rock. A competent person may determine that a protective system is not required for trenches that are less than 5 ft deep. Trenches 20 feet deep or greater require that the protective system be designed by a registered professional engineer.

However, a wider excavation can be considered a trench if forms or other structures are installed such that the distance from the edge of the form or structure to the side of the excavation is less than 15 ft.

Numerous precautions should be taken when excavating or working in trenches. OSHA has specific regulations (29 Code of Federal Regulations 1926, Subpart P) that govern most subsurface excavations.

Requirements for Trenches and Excavations

A complete and detailed rundown of all the rules and regulations for trench and excavation safety would be far too lengthy to tackle in a tailboard safety meeting. But the following are a few points to remember.

- Before beginning any subsurface work such as trenching, contact 811 or **clickbeforeyoudig.com** for the local utility alert service to establish the location of other underground service lines such as natural gas, sewer, telephone, electric power, and cable.

29

- Every trench must have a safe and ready means of exit. If a trench is deeper than 4 ft, a stairway, ramp, ladder, or other means of exit must be available within 25 ft of a worker in the trench.

- Don't expose workers in trenches to overhead loads handled by lifting or digging equipment.

- If it is possible that an oxygen deficiency or hazardous atmosphere may exist in a trench or excavation, the air in the excavation must be tested before employees enter and while work is being conducted. If necessary, adequate ventilation must be provided.

- If hazardous conditions exist (or may exist), emergency rescue equipment, including a breathing apparatus, safety harness and line, and basket stretcher must be readily available near the trench.

Unless the excavation is made in stable rock, any trench greater than 5 ft in depth must be inspected by a qualified person and if conditions warrant, a protective system (such as shoring) must be installed.

Source: omphoto/Shutterstock.com

For more information, go to the OSHA website on the topic: **www.osha.gov/ trenching-excavation**

Quick Equipment Checks

Because of a concern for the safety of you and your family, you probably periodically conduct a safety inspection of your car, looking at things such as tire wear and working brake lights. But do you do the same type of inspection on the job?

Jobsite inspections can effectively reduce workplace accidents. Unfortunately, we often neglect to keep a close watch for similar flaws in our tools and equipment that might give us an advanced warning of a hazardous condition.

Fiber rope is a much used, and often abused, tool that is seldom inspected for flaws. Fiber rope damage, wear, and strand failure often occur beneath the surface and can often only be detected by a visual inspection of unraveled strands.

Wire rope slings also require regular inspection because the first signs of failure often are not readily noticeable. A rope failure could result in a crippling injury or even death.

Safety checks of tools and equipment should be a regular part of the daily job routine. The inspections do not need to be a time-consuming chore, but they need to be done to maintain safety.

The following are other work items you should regularly inspect.

- Tool handles: Look for splinters, splits, and loose metal parts.
- Air hose fittings: Evaluate their condition and security.
- Pipe wrench jaws: Are they worn out?
- Vibrating-type air tools: Look for cracks, flaws, or other failures.
- Chains used for hoisting or pulling: Look for cracks, wear, link elongation, or deformed hoods.

OSHA has developed a hand and power tools guide that can be adapted for the needs of your utility or jobsite. This guide covers topics such as safeguards, mechanical hazards, nonmechanical hazards, protective equipment, machinery maintenance, and electric hazards. Or customize your own to identify areas of concern or those in need of attention or corrective action before any maintenance situation becomes an emergency.

For additional information, go to: **www.osha.gov/Publications/osha3080.pdf**

Jackhammer Safety

One of the most powerful tools used in the water utility industry is the jackhammer. Jackhammers are designed to break asphalt, concrete, and rocks. They come in either electric or pneumatic models. Without proper training and PPE, workers can inflict serious injury to their feet and other parts of the body, as well as injure others nearby, while operating this tool.

Before Operation

- Always wear proper PPE, which includes eye protection; long-sleeved clothing; sturdy, full-length pants; steel-toe boots or shoes; respiratory, head, and hearing protection; and safety gloves.

- Know how to safely operate the supply compressor—especially in emergencies.

- Place the compressor as far as possible from the work area to reduce the level of noise.

- Regularly inspect the jackhammer and other necessary tools for defects or damage.

- Check if all components are complete, securely in place (or tightened), and in good condition. Do this before every shift or start of operations.

- Check air hoses for breaks, cracks, and worn or damaged couplings.

- Ensure that the rating of the hose is sufficient for the job intended.

- Inspect the electrical cord for frays, wear, and other signs of damage.

- Inspect the tool's breaking point. Never use a broken or cracked point.

During Operation

- Sling the electrical cord onto your shoulder when in use to prevent the cord from accidentally swerving, which can cause electrocution.

- Always use the proper weight jackhammer for the job. For your back's sake, try to use a lighter jackhammer for the job as much as possible.

- Always lift the tool jackhammer properly, using your legs. This method helps to prevent back strain or injury.

- Use the proper jackhammer point for the material to be broken, e.g., rock point for rocks; spade point for asphalt; chisel point for concrete.

- When moving the jackhammer from place to place during operation, place your hand between the handle and the operating lever.

- Always operate the tool at a slight angle while leaning it back toward you. This way, you prevent the point from getting stuck in the material and the tool from getting out of control.

- Shut off the air supply and relieve pressure from the supply hose before changing tool points. Do the same when leaving the jackhammer unattended.

- Immediately remove defective or malfunctioning jackhammers and other tools until they are properly repaired.

- Barricade the work area as much as possible to keep spectators and untrained personnel from being exposed to the hazards of jackhammer operations.

Rules on Silica Dust

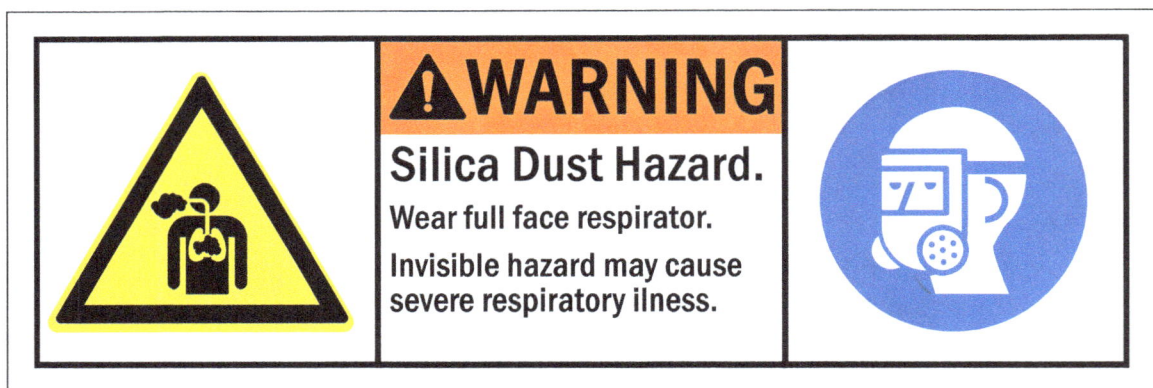

Source: gobigoo/Shutterstock.com

OSHA has proposed rulemaking for respirable crystalline silica, inhalation of which puts workers at risk of silicosis, lung cancer, lung disease, and kidney disease. Exposure to silica dust can occur when cutting, sawing, grinding, drilling, and crushing stone, rock, concrete, brick, block, mortar, and industrial sand (including sand blasting).

For additional information, see the OSHA booklet on hand tool safety: **www.osha.gov/Publications/osha3080.pdf**; Safety Services Company's website on jackhammer safety: **www.safetyservicescompany.com/industry-category/construction-safety-using-handling-and-maintaining-jackhammers**, or the OSHA website on Crystalline Silica Rulemaking: **www.osha.gov/silica-crystalline**

An Open-and-Shut Case for Gate Valve Safety

Water service often must be turned off temporarily while emergency repairs or routine maintenance are performed on a distribution system. Because of the system's location, such work often requires traffic control measures to be conducted safely. You should take time to familiarize yourself with state and local laws relating to work zone traffic safety and the Manual on Uniform Traffic Control Devices. Sometimes a gate valve must be manually operated to isolate the area where the work is being conducted. Manually operating gate valves can cause a variety of injuries, including sprains and strains of the back, knee, shoulder, elbow, and wrist. Some safety tips to keep in mind when operating a large gate valve follow.

In Traffic
- Use warning lights and flashers if you stop your service vehicle in traffic.
- If the valve is located in the middle of the road, park your vehicle between the valve and oncoming traffic.
- Use traffic cones to mark your vehicle and work area to help protect you from oncoming traffic.
- Wear appropriate protective equipment, which may include a hard hat, steel-toe safety shoes, work gloves, and an ANSI/ISEA Type "R" Class 2 or 3 high-visibility retroreflective roadway safety garment.
- For more information: **www.osha.gov/highway-workzones**

Operating the Valve
- Remove the gate lid with a pry bar or other appropriate tool.
- Use a valve key that is the correct size and length. You may have to use a key extension to get the proper length.
- Make sure the key fits tightly on the valve nut. Watch out for rounded or spalled nuts.

- When you are operating the valve, the key should be at chest level. Do not use a key that is too long (above your shoulders) or too short (below your waist).

- Know the proper direction for opening and closing the valve. Some valves are left-hand turn.

- Grip the valve key firmly with both hands when you turn it.

- When operating the valve, maintain good footing, with your feet at least shoulder-width apart.

- Position your body as close to the valve key as possible.

- Turn the valve key with slow, controlled movements. Bend your knees if necessary.

- If the valve becomes too difficult to turn, ask another worker to help you, or use a valve-operating machine.

- Don't leave the key on the valve unattended because it may present a hazard for vehicles or pedestrians or provide unwarranted access to the water system.

- Secure the gate lid when service is completed.

Additional Notes

When using handheld valve operators, follow the safety guidance in the operator's manual.

Remember to follow the OSHA Hazardous Energy Control Lockout/Tagout requirements: **www.osha.gov/control-hazardous-energy**

Biohazards and Worker Safety

What are biohazards? Biohazards are materials or human waste that cause infections or disease. Bloodborne pathogens, human waste, and drug paraphernalia are also considered biohazards and can pose a significant health threat. If you work in areas with wastewater, medical waste, or live sanitary sewer lines, for example, you should assume that all surfaces are contaminated (have germs).

Following are five sources of biohazard risk to human health:

- bacteria (e.g., *E. coli* and *Salmonella*),
- fungi (e.g., mold and yeast),
- viruses (e.g., hepatitis, HIV),
- pathogens (e.g., *Giardia* and *Cryptosporidium*), and
- endotoxins (from decaying debris).

Four ways in which the human body can be affected by a biohazard are

- ingestion (eating, swallowing),
- inhalation (breathing or smelling),
- contact (through broken skin or mucous membrane), and
- injection (stuck with a sharp object such as a needle).

Workplace Preparedness

If your work typically brings you into close proximity to biohazardous materials, you likely already know the potential safety and environmental risks and the safe handling procedures. But it's essential that everyone in the area knows what to do in a biohazard emergency, both during the emergency and afterward, during cleanup.

A properly outfitted work area contains a safety shower, an eye wash station, and a hand-washing sink as permanent fixtures. There should also be at least one well-stocked biohazard spill kit containing goggles, gloves, shoe covers, breathing masks, biohazard waste bags, disinfectants, sharp-instrument containers, and instruments for picking up sharp tools or objects such as broken glass. The kit should also contain absorbent material designed specifically for handling common biohazards, such as blood.

Be sure everyone is familiar with the biohazard safety procedures, the contents of the spill kit, the instructions for using the kit, and any SDS that may be included.

Recognizing the Threat

Most people don't know what type of condition is considered a biohazard and are unprepared to safely deal with biohazards.

Let's say, for example, that a coworker receives a serious cut while on the job. Exposure to the blood from that cut could be a problem because in the general population, one in 300 people are HIV positive; one in 20 have hepatitis; one in five have herpes; and one in three have some type of bloodborne disease, according to the CDC. What's more, the CDC says hepatitis B virus can survive for at least one week in dried blood. The virus may survive on environmental surfaces, contaminated needles, and/or instruments.

Diseases from air- and blood-borne pathogens or feces are spread most often to humans during cleanup because of improper safety equipment. For example, hantavirus is transmitted by infected rodents. Individuals become infected with hantavirus by breathing aerosolized urine, droppings, saliva, or nesting materials. A specialized respiratory mask (one that filters viruses) should be used when removing suspected nesting areas and rodent feces.

Proper Cleanup Procedures

It is especially important to adhere closely to the biohazard cleanup laws. They are imposed by multiple agencies to protect the public's health and safety. OSHA is one of the agencies that sets standards in biohazard cleanup laws. According to OSHA, "Personnel associated with the biological cleanup must be trained, immunized, and properly equipped to do so."

Biohazard restoration includes cleaning not only the visible but also the invisible. The standard for cleaning and restoration of biohazards is set by the American Bio Recovery Association: **www.americanbiorecovery.org**. As a general rule, for any blood or fluids, all visible areas should be cleaned, including all materials surrounding the affected area. When it comes to porous materials such as drywall, sometimes replacement is necessary. Cleaning of biohazard areas should include all surfaces—walls, ceilings, carpets, flooring, fixtures, switches, railings, and trim—using chemicals produced specifically to kill microorganisms.

Source: Pressmaster/Shutterstock.com

Disposing of biohazard materials after cleanup is regulated by the USEPA, OSHA, and state and local governments. All of the guidelines and regulations are written with the specific intent of lowering your infection risks and keeping you from contracting or spreading disease.

For more information, see the following:

- OSHA website on biological agents, **www.osha.gov/biological-agents**
- CDC's website on biosafety: **www.cdc.gov/safelabs/resources-tools/biosafety-resources-and-tools.html**
- The National Institute for Occupational Safety and Health (NIOSH): **www.cdc.gov/niosh/docs/2022-129**
- AWWA: *Environmental Compliance Guidebook: Beyond US Water Quality Regulations*, **http://store.awwa.org/product/1549**

Weld Well to End Well

The American Welding Society has identified more than 80 types of welding and allied processes in commercial use. Some of the more common types include oxygen–acetylene, gas–metal, gas–tungsten arc welding, shielded-metal arc welding, resistance welding, and brazing. Welding and cutting are not without risk and may lead to eye and skin injuries, respiratory hazards, electric shock, and fire in confined spaces.

Eye Injuries

Welding and cutting operations are a major source of eye injury. Related accidents occur when proper PPE is not worn. The most common eye injuries result from flash burn, metal flying into the eye, and particulates falling into the eye. The only measure that will prevent eye injury is the use of appropriate PPE. It is also important not to wear contact lenses while welding or near where welding is taking place.

The welder also must be concerned about the effects of the welding operation on nearby personnel and should always use a welding curtain or wall.

Skin Injuries

Injuries to the skin usually result from ultraviolet rays or from hot metal. The hot metal may be the material being worked on, or it may be part of the equipment.

Unprotected skin is at risk for injury. In addition to burns, it is easy for exposed skin to be cut during work with sharp metal. Proper safety shoes, clothing, and PPE will greatly reduce the chances of skin injury.

Respiratory Hazards

Without adequate ventilation or when adequate PPE is not used, the threat of respiratory injury greatly increases. Before welding, the welder should know what the metal is and the potential effects of the fumes produced, which include carbon monoxide.

Inhaling welding fumes or gas can produce metal-fume fever, the symptoms of which include a dry, metallic taste in the mouth; fatigue; nausea; and muscle and joint pain. Depending on the metal or alloy, the results can be fatal.

Adequate ventilation (natural, mechanical, or respiratory) must be provided for all welding, cutting, brazing, and related operations. Adequate ventilation means enough ventilation so that a person's exposure to hazardous concentrations of airborne contaminants is maintained below the level set by federal standards.

Electric Shock

Whenever electricity is used, a potential for electric shock exists. Only trained personnel should operate welding equipment. Be sure equipment is properly installed, inspected, operated, and maintained. Equipment should be inspected before every use. Consider the following:

- placement of welding machines,
- placement of cables,
- load protection, and
- use of electrodes and holders.

Always be aware of the potential for electric shock when welding.

Fire Hazard

Welding and cutting should be done in designated areas that are free of flammable materials or conditions favorable to fire or explosion. If your utility has a hot-work permit program, make sure to follow its requirements. Before and during the welding operation, the welder and safety watch should

- inspect the area for flammable and combustible material before welding or cutting begins,
- cover cracks or floor openings, and
- have fire extinguishers on hand.

During welding, constantly watch for fires between walls, on opposite sides of metal partitions, or in any concealed area.

Confined Spaces

Because of the small size and questionable atmosphere in most confined spaces, welding and cutting in such spaces require very serious thought and planning. The safety regulations dealing with welding and cutting in confined spaces should be reviewed.

For additional information, go to the OSHA Safety and Health Topic web page: **www.osha.gov/welding-cutting-brazing**

You've Got the Power! Of Power Tools

Hand tools are part of a utility worker's daily work. But as we all know, they can be quite dangerous. One of the main safety precautions workers can take to ensure their safety when using power tools is to understand and control the power source of power tools.

Take the following precautions and be sure to inspect your tools so that you and your coworkers can stay safe when using power tools.

Source: Billion Photos/Shutterstock.com

Work Area

Safety begins with a clean, uncluttered, and well-lit work area. Never operate power tools near flammable liquids, gases, or dust as sparks may ignite them.

Grounding

Grounded three-prong tools should be plugged into a grounded electrical outlet. In damp locations, power tools should be plugged into a ground fault circuit interrupter outlet. Never leave, store, or use power tools in wet conditions.

Transporting Power Tools

Do not carry a tool by its cord or pull the cord to unplug it from an outlet. Examine the tool's power cord before using it and protect it from anything that might cut or melt it. Hold power tools by the insulated handles or grips to avoid possible electrical shock.

Using Extension Cords

When using an extension cord, make sure it is in good condition. If it is used outside, it should be marked with a 'W' or 'W-A' as this designates it for outdoor use.

Extension cords should be compatible with the amps of the power tool you are using. Tools should be marked in a manner that specifies the length and wire gauge needed for that tool.

Pneumatic Power Tools

Some hand tools are powered by compressed air. These pneumatic tools include nail guns, staplers, drills, sanders, and others. Dangers of using these tools include poor connections between the tool and the air line, as well as the projectiles of some pneumatic tools, such as nail guns.

Use a positive locking device or safety wire on the connection between the tool and the air hose. Take the same precautions with air hoses that you do with electrical cords.

Be sure the tool is equipped with a safety tip that prevents the tool from ejecting nails or staples unless it is in contact with a solid surface.

Gas Powered Hand Tools

Gas powered hand tools present fuel vapor and carbon monoxide hazards. Be very careful operating these tools near combustible materials. Never refuel near combustibles, and always allow the motor to cool down first.

Powder Actuated Power Tools

Some utilities use powder actuated power tools. The use of these tools requires training and a special license.

If you are trained and licensed to operate a tool such as a powder actuated nail gun, be sure to inspect it to make sure the barrel is clean and unobstructed, the tool has the proper guards and shields, and that you are not in a flammable or explosive atmosphere.

Clean and lubricate all power tools only as directed by the operating manual as some cleaning agents may damage wiring or plastic tool parts.

Storage and Service

Store all tools in protected areas with all power sources detached. If safety labels or information plates become unreadable, replace them.

Make sure all power tools are serviced only by qualified repair personnel.

For more information on using power tools safely, review "Power Tool Safety Tips from OSHA" at **http://ohsonline.com/articles/2010/03/19/power-tool-safety-tips-from-osha.aspx** and **http://choosehandsafety.org**

Facing Up to Stress

Are you experiencing stress? Surveys and research reveal that

- an estimated 75%–90% of all visits to primary care physicians are for stress-related

- complaints or disorders;

- more than 40% of all adults suffer from stress-related adverse health effects; and

- stress has been linked to all the leading causes of premature mortality, including heart disease, cancer, respiratory ailments, accidents, cirrhosis, and suicide.

Stress is a normal reaction to the demands of life. You may feel stress from many events, some happy and joyous—a new job, relocation, marriage, or the birth of a child—or somber events, such as divorce or a death in the family. Even holidays or buying a new car can cause stress. Everyone feels stress from time to time.

Everyone responds differently to stress. What one person ignores or finds challenging may cause stress in another. So, do you suffer from stress?

Symptoms

Some of the most common signs and symptoms of stress are

- constant fatigue;

- muscle tightness or tension;

- anxiety;

- indigestion;

- nervousness or trembling;

- insomnia;

- loss or increase in appetite;

- grinding of teeth or jaws; and

- general complaints such as weakness, dizziness, headache, stomachache, or back pain.

Many of these symptoms may be caused by other health problems, such as the flu, but if you have one or more of these symptoms that last longer than a week, talk to your physician. You may be suffering from stress.

Reducing Stress

So, you're under stress. How can you learn to reduce the stress or control its negative consequences? Here are a few simple tips that can help reduce or control stress.

- Identify the causes of stress in your life.
- Share your thoughts and feelings with someone you trust.
- Avoid sad thoughts; try not to get depressed.
- Simplify your life as much as possible.
- Learn to manage your time effectively.
- Limit alcohol intake and avoid using illegal drugs or prescription drugs in ways other than prescribed.
- Exercise regularly.
- Practice relaxation techniques, such as deep breathing, stretching or meditation.
- Make time to unwind and have fun.
- If necessary, seek professional help.

Many sources of help are out there. Often, just talking to a friend can help, but if that doesn't work, talk to your community-based or faith-based organizations, or a licensed therapist. In addition, many companies provide access to an employee assistance program (EAP), which can provide a wealth of confidential professional counseling resources to help you, your family, or your fellow employees through difficult or stressful periods in life.

Finally, remember: it's your life. Successfully managing stress leads to a healthier, happier, and more productive life.

Need Support Now?

If you or someone you know is struggling or in crisis, help is available. Call or text 988 or chat at **988lifeline.org**

For more information, go to Mayo Clinic's recommendations on coping with stress: **www.mayoclinic.org/healthy-lifestyle/ stress-management/basics/stress-basics/hlv-20049495**

CDC's website: **www.cdc.gov/mentalhealth/cope-with-stress**

Lightning: The Underrated Killer

An estimated 25 million lightning flashes occur each year in the United States. Over the past three decades, lightning has killed an average of 58 people per year, which is greater than the annual average for either tornadoes or hurricanes. Nearly 75% of all US lightning fatalities occur during June, July, and August, and the most incidents occur between 2 p.m. and 6 p.m. The top five states reporting lightning-caused deaths are Florida, Texas, Colorado, Alabama, and North Carolina. Because nine out of every 10 lightning casualties involve only one victim and there's typically no mass destruction, getting struck by lightning is unfortunately underrated as a safety risk.

The National Lightning Safety Institute recommends that all businesses, and especially those that typically have workers with outdoor jobs, prepare and distribute a lightning safety plan to all employees. The core of the plan is to anticipate a higher-risk situation and move to a low-risk location. These plans should be site-specific, but they all share a common outline.

- *Watch for developing thunderstorms.* Thunderstorms occur year-round. As the sun heats the air, pockets of warmer air start to rise, and dark, thick cumulus clouds form. Continued heating can cause these clouds to grow vertically into massive formations that often indicate a developing thunderstorm.

- *Seek safe shelter.* Lightning can strike as far as 10 miles from the area where it is raining. That's also about the distance from which you can hear thunder. Remember that if you can hear thunder, you are within striking distance. Seek safe shelter immediately.

- *Stop outdoor activities at first thunder.* Most lightning deaths and injuries occur in the summer. Golfers and boaters are prime moving targets for a lightning bolt. Where organized outdoor sports activities take place, coaches, camp counselors, and other adults must stop activities at the first sound of thunder to ensure everyone has time to get to a large building or enclosed vehicle. Leaders of outdoor events should have a written plan that all staff are aware of and can enforce. And never seek shelter under a tree!

- *Avoid electrically connected activities indoors.* Get inside a building. Stay off corded phones, computers, and other electrical equipment that can put you in direct contact with a surge of lightning-caused electricity. Also, stay away from pools (indoor or outdoor), tubs, showers, and other plumbing. Get surge suppressors for key electrical equipment. Install ground fault circuit interrupters on circuits near water or outdoors. When inside, wait 30 minutes after the last clap of thunder before going outside again.

- *Help a lightning strike victim.* Lightning victims do not carry an electrical charge, so they are safe to touch—and will likely need urgent medical attention. For those who die, cardiac arrest is the immediate cause. Some deaths can be prevented if the victim receives the proper first aid immediately. Call 911 and perform CPR if the person is unresponsive or not breathing.

Lightning is dangerous. With common sense, you can greatly increase your safety and the safety of others. At the first clap of thunder, go to a large building or fully enclosed vehicle, and wait 30 minutes after the last clap of thunder before you go back outside.

For additional safety information, go to the National Weather Service website: **www.weather.gov/safety/lightning-safety** or the National Lightning Safety Institute site: **http://lightningsafetycouncil.org**

Texting and Working Don't Mix

It is well documented and understood that texting while driving is extremely unsafe. The NHTSA found that drivers who use hand-held devices while driving are four times as likely to get into crashes serious enough to injure themselves or others.

What about the risks caused by mobile phone and/or smart device use while operating machinery or while on a construction site? These risks in the workplace are less documented but can have the same fatal consequences. For instance, a worker severed several fingers on one of his hands while operating a chop saw; he was holding his phone between his neck and ear when the accident occurred.

Some of the main issues presented by mobile phone and smart device use while operating machinery, using vehicles, or on a construction site are discussed here.

Source: Geber86/Shutterstock.com

Distractions

Use of mobile phones or smart devices requires cognitive, visual, and manual attention. This means that any time a worker is using one of these devices, his or her mind is not fully engaged on the job at hand. Using mobile phones can also decrease productivity.

In a workplace environment that requires a high level of self-awareness, being distracted can result in high-consequence accidents, including loss of life.

Entanglements

Mobile phones or smart devices can get entangled in machinery and interfere with proper use of PPE. Similar to restrictions on wearing jewelry, which is often not allowed in high-risk work environments for this exact reason, it is important that personnel refrain from using these devices.

If a mobile device is dropped, the employee's impulse may lead her to reach into moving machinery to retrieve it, risking injury or loss of life. A worker could also place himself in danger by removing some of his PPE, such as a hard hat in order to put his cellphone to his ear, or his safety gloves to send a text.

Distractions and entanglements are issues that workers do not want to have while completing jobs that often require both hands and always require their full attention.

Operating heavy machinery is particularly hazardous; tens of thousands of injuries related to forklifts occur every year. Many injuries happen when lift trucks are driven by distracted drivers who inadvertently drive off loading docks or into fellow coworkers or other items or tip the forklift over; some accidents happen when a distracted worker falls off an elevated pallet. Many heavy machinery jobs common on a construction site and grounds work require every person onsite to have their full attention on the task at hand.

How do you start to change the culture regarding use of mobile phones and smart devices? Create a policy that includes the following:

- a Purpose Statement that explains why it is dangerous to use such devices in a high-risk working environment;

- a limit on a broad range of devices that should not be used while working a physical job;

- who the policy applies to—explicitly state that the policy applies to not only the staff workers but also contractors, consultants, temporary workers, and all personnel affiliated with the listed parties that are on the job site; and

- a clear definition of where and when workers can and cannot use their mobile and smart devices while on the job site or using vehicles.

The Right Attitude

Even if employees recognize the dangers of using mobile devices on the job, they must commit to following the policy. They must:

- recognize situations where use of cellphones can interfere in their ability to perform tasks without injury or from completing jobs in a timely manner and

- be willing to speak up when they see coworkers putting themselves in harm's way by texting or talking on the phone while performing their job duties—and if they are on the receiving end of a text from a coworker who is performing a physical job, they should not respond in kind.

When used appropriately, mobile devices can make our lives easier and more enjoyable, but when used at the wrong time and in the wrong manner, these same devices can cause serious injury.

Lockout/Tagout: Water Under Pressure Poses Danger

Fire hydrants are not just for fire protection. Water utilities use them to flush water mains, control pressure when working on water mains, and supply potable water service in bypass situations. But when is it necessary to tag an open fire hydrant as being out of service?

A hydrant requires a visible notice when it is broken or when it is open and unattended. Verbal notifications are never sufficient. Here's an example of why.

Several water utility employees were hurt, two seriously, when a firefighter unknowingly closed an untagged hydrant. The hydrant had been left open to relieve pressure while work was done on valves in a nearby excavated pit. Two valves had been closed to isolate a section of main so water department employees could cut and plug a 4-in. service branch. They opened a hydrant to prevent pressure buildup in the isolated main. Via telephone, they notified the fire department that the hydrant would be out of service until further notice—but they failed to attach an out-of-service tag to the hydrant.

At about the same time, a nearby homeowner noticed water running from a hydrant and reported the leak to the fire department. A firefighter went to the site and saw a small stream of water running from the hydrant. So he closed it! He did not see the water department crews working in the nearby pit.

The water department employees working in the pit had just replaced the fittings on the end of the pipe and were collecting their tools when the increasing water pressure blew off the push-on fittings with a high-velocity blast of water. One worker escaped with only minor injuries. But two others suffered broken bones, lacerations, and multiple injuries to the head, neck, back, and legs.

Tagging Out Fire Hydrants

OSHA cited and fined the water department for violating the standard for controlling hazardous energy through lockout/tagout. Subsequently, the department was required to create a job hazard analysis for cutting and capping pipe and to develop an effective method of lockout/tagout to warn when a hydrant is out of service.

The water department's solution was to purchase orange out-of-service bags that cover hydrants whenever a main is being isolated and a hydrant is opened to release pressure. The utility also met with the local fire agencies to demonstrate the bags and explain their purpose to the fire crews.

OSHA defines water under pressure as a hazardous energy and requires "employers to establish a program and utilize procedures for affixing appropriate lockout devices or tagout devices to energy-isolating devices (such as hydrants) and to otherwise disable machines or equipment to prevent unexpected energization, startup, or release of stored energy in order to prevent injury to employees."

Utilities need to establish programs to teach employees about the dangers of water under pressure and to explain when a tagout device must be used.

For additional information, go to the OSHA website on controlling hazardous energy: **www.osha.gov/control-hazardous-energy**

Accident Investigation: Key to Preventing Future Accidents

Your safety program is in place, your employees have been trained, and still, accidents will happen at work. When they do, they need to be investigated, and this should be considered a vital part of your safety program. Why should accidents be investigated?

- To identify the causes of the accident

- To recommend corrective actions

- To prevent the accident from occurring again

An accident should always be investigated if it results in one or more of the following:

- fatality or fatalities,

- serious injury,

- property damage, and/or

- near miss of any of the above.

An accident investigation should be handled by the supervisor(s) involved, a safety manager or inspector if there is one on staff, and/or a safety committee consisting of various employ-ees. Anyone involved in an accident investigation should have appropriate training from a certified safety professional. The investigation should have the following components.

Planning
- Accident reporting policy in place

- Investigation training

- Development of report forms

Fact Finding
- At the scene

- Time of day, location, and type of work being done

- Safety protection devices provided; were they being used?

- Actions that caused the accident

- Preservation of evidence

- Interviewing of witnesses (Who? What? Where? When? Why? How?)

- Collection of evidence, including photographs

Analyses
- Review of data

- Distinguishing facts from opinions

Conclusions
- Each identified contributing factor should be addressed

Recommendations
- One for each conclusion

- List of corrective actions

- Follow-up

Implementation of Corrective Actions

The form on the next page will help you in your accident investigation. Make sure that all employees are aware of the program. Ask employees, when they are on a job-site, to be aware of their surroundings and the actions taking place. Their awareness could help in an accident investigation and maybe even prevent the accident from happening in the first place.

For more information, read AWWA Manual of Practice M3, *Safety Management for Utilities*, eighth edition.

Accident Investigation Form

Date/Time: _____ Person(s) injured:

Describe injury:

Describe property damage:

Describe activities or job being performed:

Cause Factors

Y N PROCEDURES
- ❑ ❑ Are procedures established?
- ❑ ❑ Are procedures written?
- ❑ ❑ Was employee familiar with procedure?
- ❑ ❑ Was supervisor familiar with procedure?
- ❑ ❑ Were procedures followed?

Y N EQUIPMENT and TOOLS
- ❑ ❑ Was the proper equipment used?
- ❑ ❑ Was the proper equipment available?
- ❑ ❑ Was the proper equipment on site?
- ❑ ❑ Was the equipment properly maintained?
- ❑ ❑ Was the employee trained to operate the equipment?
- ❑ ❑ Were the proper tools used?
- ❑ ❑ Were the proper tools available?
- ❑ ❑ Were the proper tools on site?
- ❑ ❑ Were the tools properly maintained?

Y N TRAINING/EXPERIENCE
- ❑ ❑ Was employee trained for task?
- ❑ ❑ Was training documented?

Y N LIFTING
- ❑ ❑ Was the item under other equipment?
- ❑ ❑ Was the item stuck?
- ❑ ❑ Was the item too heavy?
- ❑ ❑ Was item in awkward position?
- ❑ ❑ Was help requested/received?
- ❑ ❑ Was help available?

Y N PERSONAL PROTECTIVE EQUIPMENT
- ❑ ❑ Was PPE available?
- ❑ ❑ Was PPE used?
- ❑ ❑ Was PPE appropriate for the job?
- ❑ ❑ Was PPE properly maintained?
- ❑ ❑ Were respirators used?
- ❑ ❑ Were employees trained in use of PPE?

Y N SUPERVISION
- ❑ ❑ Was supervisor at site?
- ❑ ❑ Was employee deficient in skill or ability?
- ❑ ❑ Has employee accomplished this task before?
- ❑ ❑ Employee had ( ) mo/yr experience?

Y N OTHER FACTORS
- ❑ ❑ Were allergies, hearing, eyesight, or inadequate strength factors?
- ❑ ❑ Was fatigue a factor (overtime or second job)?
- ❑ ❑ Did employee suffer heat exhaustion?
- ❑ ❑ Was stress a factor (job or other)?
- ❑ ❑ Was lockout/tagout performed?

Reagents in Disguise:
Chemical Safety

Chemicals should never be in disguise in a water treatment plant. Chemical reagents are substances that take part in, or bring about, chemical reactions. A laboratory technician or water operator should know how to safely label, store, and handle chemical reagents so everyone working with them knows how to stay safe.

Properly Labeling Chemicals
Reagent labeling must include:

- Chemical name,
- concentrations,
- For chemical mixtures, the proportions of each ingredient should be listed on the label,
- the date they were made, and
- the name of the lab worker (or company) who made them. (This information should also be recorded in a reagent logbook.) Other considerations when labeling reagents are as follows.
- The lab supervisor should take part in the labeling of these products. That way, information is shared with other lab technicians.
- Have a chemical incompatibility chart available when working with chemical reagents. This chart lists the chemical on one side and other chemicals and materials that should be avoided when working with that chemical on the other side.

Storing and Transportation Considerations
- Overloading or cramming lots of chemicals into an area is asking for trouble.
- Store lab chemicals in a cool, well-ventilated area out of sunlight and protected from any ignition sources.

- Lab chemicals should be stored in appropriate containers, and only compatible chemicals should be in each storage area.

- Special storage areas and procedures may be required for some chemicals, so check the SDS.

- Some chemicals, such as flammable and biohazardous substances, need to be locked up.

- Many labs require that written records are maintained when a controlled or dangerous substance is moved into or out of a storage area.

- Make sure labels and warnings are easily accessible.

- Incidents involving tanks usually occur due to damage to the valve, pressure-relief device, and/or handwheel—so never grasp these to lift or carry a tank.

- Don't roll tanks to move them; use a dolly or cylinder cart.

Disposal

- Laboratory chemicals need to be properly disposed of. They cannot simply be poured down the drain at the end of an experiment or procedure.

- Consult the SDS and lab supervisor. Chemicals can cause damage to the drains, react with other chemicals, and harm the environment.

- Only water-soluble chemicals can be poured down a drain, and most need to be diluted.

- Labs will have specified disposal procedures for many substances such as labeling them and placing them in an area for a waste disposal contractor to pick up.

- Disposal practices should be well defined, and regulatory requirements should be posted if applicable.

- If you ever discover a substance or chemical that is not clearly marked or accounted for in your list of lab chemicals, turn it over to your supervisor immediately for proper disposal.

For more information on safe chemical handling, visit the "Chemical Hazards and Toxic Substances" page of the OSHA website at **www.osha.gov/SLTC/hazardoustoxicsubstances**

Be Kind to Your Body:
Stretch Before Working

Utility work can be a physically demanding job. It frequently requires some workers to spend considerable time in awkward postures. Athletes need to warm up before the start of a workout or competition, and so do utility workers. This includes office staff as well as field workers! Through stretching you can prepare your muscles to handle the load and possibly prevent the more frequent forms of work injury: sprains and strains.

Before the start of your shift, or before heading out to the field, take a few moments to stretch. A few simple movements help increase circulation and reduce fatigue—plus you might even become more relaxed! A stretch break any time during the day will also help you feel better and work better.

Why Stretch?

A flexible body is crucial for physical activity—whether it's for sports or for work. Stretching increases flexibility, minimizes the chances of pulling or tearing muscles, and improves performance. A flexible muscle can react and contract faster, and with more force. Flexibility also increases agility and balance.

Following are a few tips to help you get the most out of stretching and exercise.

- Start out easy. If you haven't been regularly exercising, don't try to do too much in the beginning.

- Stretch regularly. Make it a routine at the beginning of every work shift.

- The warm-up should not be painful, but you should definitely feel the stretching and the working of all the muscles and joints.

- Hold each stretch for 10 seconds. Do not bounce. Breathe normally during the stretch.

Easy stretching exercises include the following.

Source: Mangostar/Shutterstock.com

- Neck rotation: Turn your head to the side, stretching your chin toward your shoulder. Turn head back to center and repeat to the other side. Increase the range of the stretch by dropping the opposite shoulder. See if you can lower your head further.

- Shoulder stretch: Stand with feet shoulder-width apart. Raise one arm overhead and stretch as far as you can without bending the torso. Repeat with opposite arm.

- Forearm stretch: Extend your right arm straight out in front of you, palm downward. With the left hand, grasp the fingers of the right hand and pull back gently, stretching the wrist and forearm. Repeat with the left arm.

- Triceps stretch: Raise one arm straight up, so your upper arm is near your ear. Bend your arm at the elbow and let your hand fall to the back of your neck. With the other arm, reach behind your head and place your hand on top of the bent elbow. Gently pull down and back on the elbow. Repeat with other arm.

- Trunk stretch: Stand with your feet a little more than shoulder-width apart. Reach your left arm overhead and bend to the right at the waist. Repeat on the opposite side.

- Torso twist: Stand at arm's length from the wall, with the wall at your side. Reach one arm out and place your hand on the wall. Reach the other arm around the body, stretching the hand to the wall. Repeat on opposite side.

Pain and discomfort probably mean you did too much. Back off a little, and if pain persists, check with your doctor.

For Prevention of Musculoskeletal Disorders in the Workplace, go to **www.osha.gov/ ergonomics/control-hazards**

Build a Construction Site Safely

Whether you are a utility field worker, inspector, manager, or supervisor, sooner or later you will work at or visit a busy construction site. When you do, keep in mind that a construction site can be the most hazardous environment in which you will ever work.

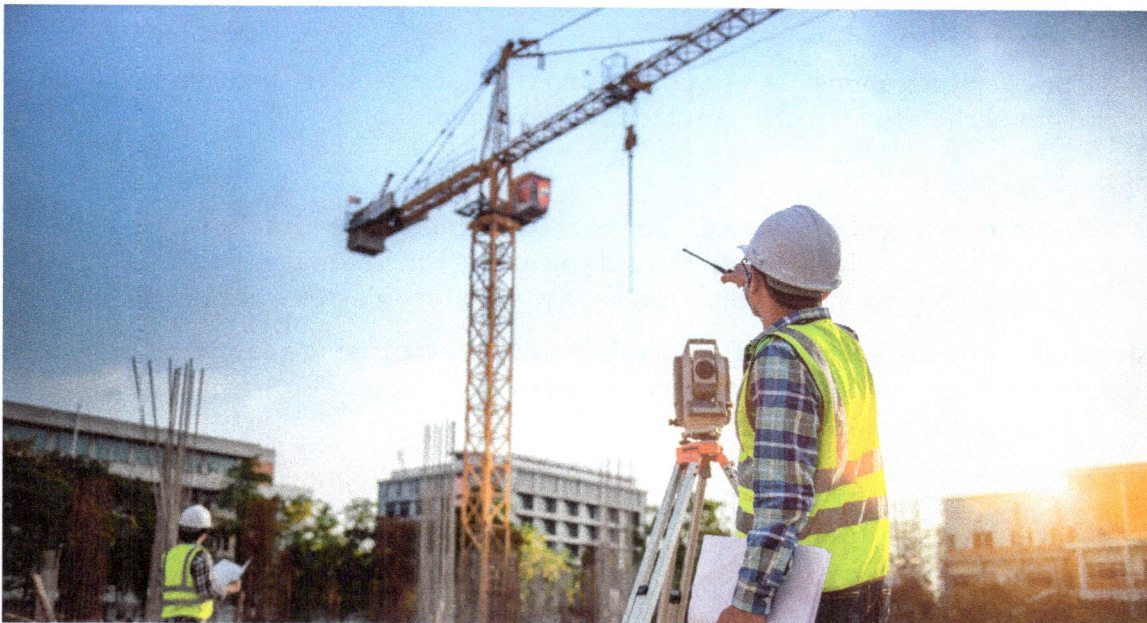

Source: SORN340 Studio Images/Shutterstock.com

Typical hazards include
- heavy equipment,
- high traffic areas, and
- flying debris.
- Potential for excessive noise level exposure

Heavy Equipment Movement

On any construction site, you are likely to find heavy equipment such as backhoes, front-end loaders, and dump trucks. The best way to prevent injuries from these large machines is to keep your distance. However, when you can't keep a safe distance, remember the following simple rules.

- Make eye contact with the equipment operator.
- Listen for backup alarms.
- Watch out for pinch points and dump trucks.
- Always have an escape route in mind.

Traffic Movement

Most construction sites make it a priority to safely move heavy equipment traffic around the site. When you enter a site, talk to the project manager about equipment movement and then mark your work area with plenty of cones, signs, and flashing arrows. If possible, park a vehicle between you and the rest of the construction site. As an added precaution, point the wheels of your vehicle in the direction you want it to roll if it is struck.

Make one person responsible for maintaining traffic control. Truck drivers who move in and out of the site regularly are often the best candidates.

Flying Debris

Watch out for flying debris such as sparks, metal scraps, hot hydraulic fluids, dirt, and rocks. These can be launched toward you at any time, so make it a habit to be vigilant of your surroundings and to always wear safety glasses when onsite.

Maintain a safe distance from flammable materials when using a saw or grinder, and position your work so sparks fly away from flammables. But also watch out for how sparks may affect coworkers.

Watch for small pieces of metal flying off flaring tools or hammered pieces of steel. To prevent this hazard, grind off any burrs on the piece of metal being worked. Broken hydraulic hoses on heavy equipment can expel hot hydraulic fluid (another reason to keep your distance from heavy equipment).

Keep the Site Neat

What is true around the office or at home is also true on a construction site—a little housekeeping can go a long way toward creating a safer environment. Here are some recommendations.

- Keep the construction site as clean as possible. Pick up discarded scrap materials and debris, including wood, protruding nails, forms, and fasteners. Work areas, passageways, and stairs should especially be kept clear and free of debris.

- Provide separate waste containers for construction debris, office waste, and trash or garbage.

- Provide an appropriate container, with a lid, for hazardous wastes such as oily rags and flammable solvents.

- Keep incompatible materials separated.

For more information, go to the OSHA Pocket Guide on Construction Safety: www.osha.gov/Publications/OSHA3252/3252.html.

Don't Be Shocked by Charged Pipes!

According to an AWWA study, more than 350 significant distribution system electric shock incidents occur annually to water utility workers. A much larger number of minor shock incidents occur each year, many of which go unreported.

Electric shock is a danger water utility workers face during the installation and repair of water pipes and meters. Water pipes are often used to ground electricity in homes. If there is a fault in the electric system, the pipe or meter can be energized with electricity. A severe or even fatal shock can occur if enough electricity is present in the pipe or meter.

Some utilities insulate the water service at the corporation stop or meter. Electrical insulation of water services has proven to be very effective in reducing the number of shock incidents. However, many uninsulated services remain. So what steps should utility workers take to avoid being shocked on the job?

Understand the Hazard

Electricity always wants to return to its source to complete a continuous circuit. A typical circuit has two conductors: one that flows from a service panel to an appliance and one that returns the current to the panel. A neutral wire and ground wire are both connected to electrical ground; the neutral wire completes the electric circuit by conducting current away from the plugged-in electrical device. The ground wire is a safety device that carries electric current away from a device when the circuit or plugged-in device malfunctions.

Grounding wires are connected to all outlets and metal boxes and then down to the earth by attaching it to either a metallic rod or a water pipe. The shock to utility workers occurs when they install or remove a water meter or cut through metallic pipes connected to a faulty system.

Because electricity may take multiple paths to ground, the worker may get shocked when first touching a pipe or service meter. A worker may not get shocked when removing a meter or pipe because it breaks the circuit, but he or she may be shocked when reinstalling that meter or service line because that action completes the ground circuit.

Use Proper Procedures and Safety Equipment

Every case will be a little different, but here are some general guidelines to consider when approaching meters or pipes that are part of a home's or building's ground system.

- Identify the composition of the service line to be worked on and that of nearby properties. This will help determine the likelihood of a shock hazard, because metallic water lines allow an electrical current to travel from a neighboring property. The pipes most likely to act as an electrical conductor are ductile iron, copper, cast iron, steel, and galvanized.

- Voltage-rated rubber gloves with leather "glove keepers" worn over them provide the best protection for workers and should be worn when inspecting, installing, or removing a meter, or when cutting and repairing a service line. Class 00 rubber electric-safety gloves are rated for maximum use voltage of 500 volts AC and protect against most common shock hazards associated with residential electrical systems. Consult with a voltage-rated glove manufacturer to determine the appropriate class of gloves for your utility's situation. Both pairs of gloves should be inspected prior to use and need to be tested and recertified periodically.

- Using voltage-rated gloves, check for current with a clamp-on ampere meter. The presence of amperage indicates a potential electrical problem and shock hazard. If there is evidence of an electrical problem, notify the building occupant and/or local power company so that they can determine the source and eliminate the hazard. Be aware that a zero reading does not guarantee safety, as the source of the current may not be constant (i.e., a garage door opener) and safety equipment should still be used.

- A voltage-rated jumper or bridging conductor can be used to maintain grounding or bonding capability of a pipe during repairs by connecting around it during the repair. Using voltage-rated gloves, use an emery cloth or another method to clean the pipe to bare metal. Connect the jumper, mainline side first, securely to the pipe. Jumpers with alligator clips should not be used. If current is present, the amp meter should be used to measure that current is passing through the jumper prior to removing the meter or cutting a service line. Because electricity can take multiple paths to ground, a jumper should not be used as the only protection, voltage-rated gloves should be worn during the repair. When removing the jumper, disconnect the customer side first. Voltage-rated jumpers must be inspected prior to use and need to be tested and recertified periodically.

- If a worker is shocked, he or she should seek immediate medical attention. Be aware that an electrical injury can cause arrhythmia that can be fatal hours after contact.

Some local codes now prohibit the use of water pipe grounding, but many do not, so the practice and associated hazard are still widespread. Don't take chances: follow these safety tips when working with metal pipe and meters!

Know Your Colors, and Call Before You Dig

There's a good reason why utility workers call before they dig. It protects the public and the water utility crew who will be involved in the excavation. No shortcuts can be taken when digging excavations. How often do digging-related accidents occur? According to Common Ground Alliance, 700,000 underground lines are struck each year. Aside from the risk of injury or worse, you or your crew might be held liable for lost services.

Call Before You Dig

Avoiding utility lines starts with knowing specific colors for each utility service. Some may be only feet or even inches from each other.

Lines marked on and flags placed in the ground indicate that various utilities have been to the scene and marked their lines. The colors associated with the utilities are as follows.

- White is used by the excavators to indicate where the excavation will take place.
- Red indicates electric power lines, cables, conduits, or lighting cables.
- Orange indicates telecommunication, alarm, or signal lines, cables, or conduits.
- Yellow indicates natural gas, oil, steam, petroleum, or flammable material.
- Green indicates sewers and drain lines.
- Purple indicates reclaimed water, irrigation, and slurry lines.
- Pink indicates temporary survey marking and unknown utilities.
- Blue indicates water lines.

Occasionally, a locator may have a false positive and will need to cover up a previous marking so that it doesn't become confused with the new one. Some crew members carry a can of black spray paint for this purpose.

It is the excavation crew's responsibility to maintain marks that may be removed during excavation. Some excavators use offset markers to clearly mark the location and direction to the utility they are referring to.

Weather and traffic can make it difficult to see where the utility markings are. In such instances, DON'T GUESS—call 811 for a re-mark! This might require a suspension of the project for 2–3 days, but ensuring workers' safety is worth it.

Emergency Repairs—When You Cannot Call 2-3 Days in Advance

Although utilities are usually given 2–3 days to mark their lines before an excavation begins, for an immediate water line repair that requires excavation, take the following steps.

1. Make sure 811 has been called and notified that an emergency excavation needs to occur. They will begin to contact utilities and have them shut off their lines.

2. Crews should shut down all valves and controls of the damaged water line.

3. Wait a reasonable amount of time before excavating. That determination is made by a supervisor who carefully considers the immediate threat to life or property.

4. Use Tolerance Zone "noninvasive" or "soft digging." Per 811, the excavation Tolerance Zone is "comprised of the width of the facility plus 18 in. on either side of the outside edge of the underground facility on a horizontal plane."

Laws vary, but recommended practices for noninvasive or soft digging within 18–24 in. of a marked line generally include the following.

- Use a blunt end shovel rather than a sharp one and/or a vacuum or suction excavator and pneumatic hand tools, if available.

- Do not use a pry bar or pickaxe.

- Use a gentle prying and loosening technique with the shovel.

- Do not stab at the ground.

- When possible, dig at an angle that is parallel with the utility line to loosen the dirt around the line.

- Support or brace lines when soil around them is removed.

- Treat all lines as if they were live.

What to Do When a Utility Line Is Hit

What happens when you are on a water utility crew that accidentally strikes a utility line in the area? That depends on what type of line is struck.

Regardless of the type of utility struck, immediately halt the excavation and shut off all mechanized equipment.

Following is a list of steps to take based on the type of utility struck.

If a fiber optic line is struck:

1. Do not look directly at the cut, as it can produce light capable of damaging your eyesight. If a gas line is struck:

    a. Evacuate and secure the area and call 911 and the gas company.

    b. Remove potential sources of ignition: leave all equipment turned off, including engines, phones, and two-way radios.

    c. Do not attempt to fill in the excavation; allow the gas to dissipate into the atmosphere.

    d. If the gas leak is on fire, let the fire burn and do not attempt to extinguish it.

2. If an electrical line is struck:

    a. Immediately call the power company so that they can deenergize the line.

    b. If you are on equipment, such as a backhoe, stay on the equipment until you are sure there is no current flowing through the machine.

    c. If you *must* get off, jump off and land on both feet (do not STEP from the machine). Do not touch the ground and the machine at the same time.

    d. Never touch or pull an injured coworker off the electrical line as you may become part of the current.

There is no such thing as "barely nicking" a utility line. Any time a utility line is struck, you must report it immediately to the utility involved and the 811 call center.

For more information about specific requirements by state, visit **www.call811.com**. In Canada, the website is **http://call811.com/811-In-Your-State/Canada**

Take a Load Off:
Tips for Safe Lifting

An improper lifting technique can lead to serious and possibly permanent back, leg, and arm pain. A poor lifting technique can cause both acute injury and serious chronic effects. Practice using the right lifting technique to help avoid these problems.

Whether you work in an office environment or in the field, you may encounter instances where heavy lifting is involved. Even if the item you are lifting is not something that is perceived to be heavy, it is always important to keep in mind the following tips as you plan to lift, move, and lower an object.

Plan the Lift Before You Start
Prior to moving the load from point A to point B, take a minute to do the following.

- Check the weight of the load by slightly tipping or pushing it.
- Ensure that the load is stable. Repack or secure the load or ask for assistance if the load is unstable.
- Ask for help or use mechanical equipment if the load is too heavy.
- Ensure that the path of travel is clear of items that might cause you to trip and fall.

Lifting
- Face the load with your feet shoulder-width apart.
- Bend your knees, not your back!
- Keep your back straight and your head up.
- Rest the load on your bent knee as you prepare to stand.
- Position the load close to your body.

Moving the Load

- Keep the load as close to your body as possible.

- Pay attention to where you are going.

- Avoid bending and twisting your back; turn with your feet when you need to change direction.

- If you can't see over the load, find another means to transport it.

- Face the direction you are walking. If you need to turn, stop and turn in small steps and then continue walking.

- Keep your eyes up. Looking slightly upward will help you maintain better position of the spine.

Lowering the Load

- Use leg muscles—never your back—when lowering the load.

- Set the load on a table or in another location that is at waist level.

- Watch your fingers when lowering the load.

General Tips When Moving Heavy Loads

- Pushing is always easier on your back than pulling.

- When pushing, keep your elbows close to your body and use your leg muscles instead of your arm and back muscles.

- Wear shoes that have good support and traction.

Be aware of the early warning signs of back strain. If you experience back pain, such as burning or shooting pain, numbness, or a tingling sensation, seek immediate medical attention.

For Prevention of Musculoskeletal Disorders in the Workplace, go to **www.osha.gov/ergonomics/control-hazards**

For additional information, go to MSD Solutions Lab Resources National Safety Council: **www.nsc.org/workplace/safety-topics/msd/resources**

Keep Trouble Out and Let Help in With Access Control

When an emergency occurs at a water facility, emergency rescue personnel must have unhindered access to respond to the situation. Medics must be able to provide first aid to injured people, and law enforcement personnel must have the ability to engage a threat (a person, device, vehicle, or event) to prevent a security breach or a dangerous situation.

Frequently, emergency responders will pull up to locked gates thinking they have the right passcode for entry, only to punch the code in the keypad and watch the gates do nothing. They may then resort to tailgating another car through to gain entry to where they are supposed to go. Otherwise, the emergency center dispatcher must recontact the original reporting party to get them to "buzz in" the responders, delaying emergency response.

Mandating Access

Emergency access control might be addressed in local ordinances, but in many communities, it is not. Many current codes were written years ago by fire authorities and do not take advantage of recent advancements in the access control industry. While some popular methods of emergency entry meet firefighters' approval, other public safety agencies may not have been consulted in the selection process.

Local ordinances should guide water professionals to the preferred emergency access method, but the absence of applicable codes should not determine whether such access is provided. If you want periodic facility patrols and quick response to emergencies by patrol officers and firefighters, access to your facility had better be easy, quick, and reliable.

Traditional Access Systems

Following are some basic methodologies emergency personnel can employ to gain entry to gated areas, each with its own strengths and drawbacks; some technologies can be combined to form hybrid applications.

Keypads. Some gates have combination locks or keypads that accept a hand-entered numerical passcode assigned to emergency crews. Many keypad systems lack audit control, as all emergency crews typically use the same code.

It is not uncommon for public safety to be locked out of a call for service because the code changed and no one told the agency. Consider, too, if a passcode were to fall into the wrong hands because it was broadcast over an insecure radio scanner: Who would be liable? What would be the potential ramifications?

Third party. With a dispatch callback procedure, telephone, or intercom system, residents, guards, or employees can remotely grant gated-area access to a third party. Access can also be granted directly by a guard at a perimeter checkpoint. Drawbacks to this system are that during off-hours, no one may be present to provide access, staffing is expensive, and public safety personnel cannot enter a facility covertly.

Locks. If a facility has manually operated gates, a key and lock may be the only option. Some local agencies require the use of a lockbox, which houses either a switch to activate the gate mechanism, another key, or an access card to open the barrier. This solution is used almost exclusively by fire departments.

Some lockboxes can be accessed by an infrared beam from devices such as a personal digital assistant, but the majority still require conventional keys. The downside to keys is accountability and the sheer number required to equip every emergency vehicle. A lost key might require rekeying all matching locks, switches, and lockboxes and replacing all existing keys—a costly proposition.

Cards. Access cards provide an audit trail of activity, as the system can associate each card with individual users or vehicles. Cards with a touch plate, embedded chip, and magnetic strip are inserted into or touched to a card reader to allow access. Proximity cards are read from a distance, which means the pass-through speed of emergency vehicles is increased because actual contact with the card reader is not required.

If a card is lost, the associated permissions can be quickly removed from the system. Cards are relatively inexpensive, and replacements can be quickly used. But as with keys, managing a card for every potential response vehicle can be expensive and an audit control nightmare.

Advanced Technology Systems

Light. Some municipalities use a traffic priority control system, where emergency vehicles in the jurisdiction are equipped with a coded infrared strobe light that preempts traffic signals during emergency responses, allowing a fire truck or police car to get a green light at controlled intersections. Similar receivers can be attached to facility gate controllers to provide emergency access to vehicles flashing the special strobe.

This solution requires each emergency vehicle to be equipped with a strobe emitter, which may prove cost-prohibitive and impractical for this limited use. Some emitters use visible light that may compromise the covert entry of responding units.

Sound. Sound-activated entry systems open a gate when an emergency vehicle gets within range of an audio sensor that detects the siren. Such systems are fairly inexpensive, are compatible with most gate operators, and are popular with fire departments.

However, although fire equipment typically rolls to calls with lights and sirens on, announcing the arrival of law officers this way may be the last thing law enforcement wants to do.

Sound-activated systems also preclude entry of officers on foot or bike and of other service providers, such as security and utility staff, who otherwise would have been provided an access card, code, or key.

Radio signal. A gate equipped with a radio receiver can be opened with a manual transmitter, an "always on" transmitter, or a radio frequency identifier. Manual transmitters require users to push a button to open a gate. This technology is used to activate garage door openers. Active transmitters require no user action; they continuously emit a radio signal that is detected by a gate receiver, which in turn activates the gate opener. Another type of transmitter is mounted on the underside of a vehicle where the signal is detected by a roadway loop similar to those used to detect cars at traffic signals.

Radio signal identification is quick (less than four seconds) and secure. Receiver range can be set from within inches of the receiver to about one-quarter of a mile away, and handheld or vehicle-mounted radios can be used to open the gate. An internal log in the receiver maintains details on what agency gained access and when, retaining the most recent transactions.

Problems here include the probable number of different access frequencies or technologies in any given jurisdiction, the compromising of receivers with matching frequencies in the event of loss or theft of a transmitter or transponder, and the possibility of an always-on transmitter inadvertently activating a gate when driven past a gated complex.

Forced entry. More of a method than a system—and certainly last on the list of emergency access options—is forced entry. Crashing fences, cutting locks, and breaching gates are proven means for public safety personnel to get where they need to go, but such tac-tics usually result in damage to facility equipment or emergency vehicles. Jumping fences puts emergency responders at risk of injury and leaves them without vehicle-mounted equipment.

System Override

What happens when there is a loss of power at your facility? Can people get out? Can people get in? Security gates should also include the ability to override the gate operator in case of a power or mechanical failure. Such systems include manually operated mechanisms and backup power supplies.

A battery backup system can automatically open a gate and then shut down the gate opera-tor until the primary power supply is restored. If the battery backup and primary power both fail, the gate operator should go into a fail-safe mode that allows a malfunctioning gate to be manually pushed open so that vehicles or people are neither locked in nor locked out. Fail-safe overrides are mandatory in many jurisdictions across the country.

If you increase your utility's security by installing gated systems, remember, too, to consider emergency access. Absent around-the-clock, onsite security staff and without proper controls, public safety response times can be unnecessarily lengthened. Examine the options and develop a comprehensive, holistic approach in cooperation with your local authorities. Remember, the safety of your employees and customers may depend on quick, simple, and reliable access to gated facilities.

For more information, see the AWWA book *Security and Emergency Planning for Water and Wastewater Utilities.*

Climb Into Confined-Space Safety

A confined space is any area with limited entry and exit that contains known or potential hazards and is not intended for continuous human occupancy. In the water community, these spaces include manholes, trenches, storage tanks, wells, vaults, tunnels, and trenches. Hazards within a confined space include

- oxygen deficiency by displacement with other gases and the introduction of nitrogen from cable pressurization;

- toxic gases from decomposing soil, chemical spills, and engine combustion exhaust (from vehicles and equipment);

- combustible or flammable vapors and gases from underground storage or piping facilities;

- moving equipment parts, structural hazards, entanglement, slips, and falls;

- temperature extremes, including atmospheric and surface;

- shifting or collapse of bulk material;

- drowning hazards;

- barrier failure resulting in a flood or release of free-flowing solids;

- uncontrolled energy, including electrical shock or water pressure;

- reduced visibility; and

- biological hazards.

Safety Equipment

Several pieces of equipment are required for safe entry into a confined space:

- work-area protection devices, such as traffic (reflective) vests, traffic cones, manhole guards, and work-area protection signs;

- manhole cover lifter and manhole hook;

- atmospheric tester for combustible gas, oxygen deficiency, and other toxics;
- power ventilator (blower);
- pump to remove water;
- first-aid kit; and
- portable fire extinguisher (dry chemical).

Atmospheric Conditions

Oxygen deficiency and buildup of hazardous gases poses one of the most common and lethal dangers in confined spaces. The air within the confined space must be tested prior to entry into the space.

Atmospheric conditions are considered unacceptable if oxygen levels are less than 19.5 percent or greater than 22.0 percent. The following levels of other hazards are unacceptable.

- A flammable gas, vapor, or mist greater than 10 percent of its lower flammable limit (LFL). LFL means the minimum concentration of the flammable material that will ignite if an ignition source is present.
- An airborne combustible dust at a concentration that obscures vision at a distance of five feet or less.
- An atmospheric concentration of a substance greater than the allowed limit in the SDS for that substance.

Testing Procedures

Prior to using an air monitor, it is important to thoroughly review and understand its user manual.

It is important to test the air at various levels in a confined space. Some gases are lighter than air and stay at the top of a space. Others are heavier and settle at the bottom of the space. Refer to the air monitors user guide to determine how long the monitor must remain at each level.

If there is a possibility of flammable gases beginning present, it is recommended that the area around the spaces opening is checked prior to opening.

Methane
Vapor Density 0.6
Lighter than Air

Air
Vapor Density 1

Chlorine
Vapor Density 2.5
Heavier than Air

Methane
(Lighter than air)

Hydrogen Sulfide
(Heavier than air)

Force-air Ventilation

Forced air ventilation of a space is necessary when atmospheric conditions of a space are unacceptable or natural ventilation is not adequate to maintain acceptable conditions.

There are two types of forced-air ventilation, positive-pressure, and negative-pressure. The picture below is an illustration of positive-pressure ventilation. Fresh air is pulled in replacing air in the space. The blower intake should be positioned to minimize the intake of toxic fumes (i.e., exhaust from passing vehicles).

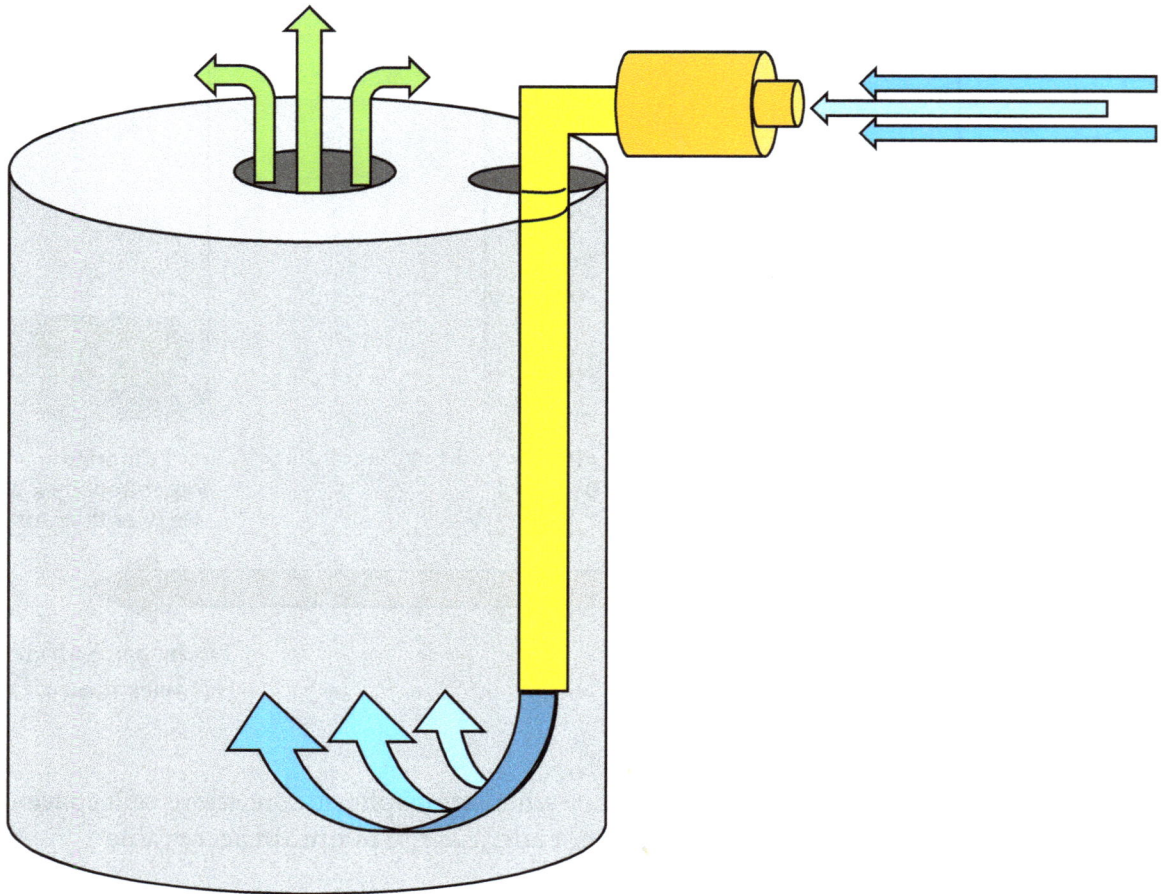

Negative-pressure ventilation pulled air from the space pulling in fresh air. When negative-pressure ventilation is used, position the exhaust so that hazardous air does not create a hazard to others.

To determine how long it will take to completely turn the air over in a space you must know the volume of the space and the flow rate (cubic feet/minute) of the fan. Divide the volume of the space by the flow rate of the fan.

For more information, go to the OSHA website: **www.osha.gov/SLTC/ confinedspaces**.

Message to Self:
Distracted Driving Is Dangerous

Distracted driving is any activity that diverts a driver's attention away from the task of driving. These distractions can be electronic, such as text messaging or using a navigation system, tablet, or cellphone, or more conventional, such as talking to a passenger or eating. Other common distractions include grooming, reading, drinking, watching a video, or changing the radio station, CD, or MP3 player. Because text messaging—texting—involves cognitive, visual, and manual attention, it has received the most legal attention in recent years, including legislative bans on texting while driving.

It's well documented in multiple university studies that drivers simply can't safely do two things at once. These studies concluded that motorists talking on a handheld or even a hands-free cellular phone are as impaired as intoxicated drivers with a blood alcohol level of .08 (the minimum level that defines drunk driving in most states).

Here are a few eye-opening statistics from the National Highway Traffic Safety Administration (NHTSA) that may encourage drivers to limit their distractions:

- Eight percent of fatal crashes, 14 percent of injury crashes, and 13 percent of all police-reported motor vehicle traffic crashes in 2021 were reported as distraction-affected crashes.

- In 2021, an estimated 362,415 people were injured in motor vehicle accidents involving a distracted driver, and 3,522 people were killed in such accidents in 2016.

- Five percent of all drivers involved in fatal traffic crashes in 2021 were reported as distracted at the time of the crashes. Seven percent of drivers 15 to 20 years old involved in fatal crashes were reported as distracted. This age group has the largest proportion of drivers who were distracted at the time of the fatal crashes.

- In 2021 there were 644 nonoccupants (pedestrians, pedal cyclists, and others) killed in distraction-affected traffic crashes.

Perhaps the most common distraction is cellphone use. About 89 percent (approximately 277 million) of Americans have a cellphone, and 77 percent of those individuals report that at least some of the time they talk on the phone while driving.

Distracted driving has risen to unprecedented levels, and state legislatures have taken action. Eleven states, the District of Columbia, and the Virgin Islands have banned handheld cell phone use for all drivers; and 41 states, the District of Columbia, and Guam have banned text messaging by all drivers.

Everyone Has a Personal Responsibility

Common sense and personal responsibility are a big part of the solution. But the problem can't simply be legislated away. Many corporations and utilities have adopted strict hands-free driving policies for their employees. President Barack Obama issued an executive order in 2011 that prohibits more than 4 million federal employees from texting behind the wheel while working or while using government vehicles and communication devices.

More portable technology is available now than ever before, and driver distractions have risen to unprecedented and alarming levels. We live in a world where people expect instant, real-time information 24 hours a day, and those desires do not stop just because people get behind the wheel. Drivers don't always realize the dangers of taking their eyes and minds off the road, their hands off the wheel, and focusing on activities other than driving.

For more information, go to the official US Government website for distracted driving: **www.distraction.gov**, or the National Safety Council website on the topic: **www.nsc.org/learn/NSC-Initiatives/Pages/distracted-driving.aspx**.

Don't Let Chemicals Get You!

Water utility operators and laboratory staff are often exposed to chemicals that can cause severe harm or even death. Many chemicals are extremely toxic, and even small quantities of them can be lethal.

The effects of chemical exposure can be local—at the point of contact—or systemic. Systemic exposure occurs when the chemical agent is absorbed into the bloodstream and distributed throughout the body, affecting one or more organs. If you are exposed to a toxic chemical, the severity of damage will depend on the toxicity of the substance, its solubility in tissue fluids, its concentration, and the duration of exposure.

A person can be exposed to dangerous chemicals in the following ways:

- Dermal contact
- Inhalation
- Ingestion
- Ocular exposure
- Injection

Dermal Contact

Spills and splashes in the laboratory or when loading chemicals into vats or mixing bays can result in contamination of exposed skin. When chemicals come in contact with the skin or the mucous membranes, they can cause surface irritation at best. At worst, the chemicals can be absorbed into the bloodstream, causing systemic poisoning. Chemicals primarily penetrate the skin through hair follicles, sebaceous glands, sweat glands, and cuts or abrasions. Touching contaminated hands to the mouth, nose, and eyes can also cause chemicals to be absorbed into the body.

Inhalation

Inhalation is the most common road of entry for toxic substances. Toxic vapors, mists, gases, and even dust and particulates can be absorbed through the mucous membranes of the mouth and nose and subsequently travel into the throat and lungs and cause serious damage to those tissues. The effects are further compounded if the substances pass through the lungs into the circulatory system.

Ingestion

Mouth pipetting in the laboratory can lead to the ingestion of chemicals, but an even more common cause of unintentional ingestion of toxic substances is from foods that were stored in containers, such as beverage jars, that had been used to store nonfood items (paint, plant food, or other substances). Another unsafe but common practice that can lead to ingestion is storing food in a place where chemicals are stored or storing chemicals in a refrigerator used for food.

Ocular Exposure

Unprotected eyes can become contaminated by splashing, aerosol contamination, or rubbing with contaminated hands. Many chemicals are capable of causing burns and loss of vision. Absorption into the bloodstream from ocular exposure can also occur quickly, because eyes contain many blood vessels.

Injection

Inattentive laboratory workers can have accidents with needles; an accidental stick can inject chemicals into someone inadvertently. Broken glass containers that contained toxic chemicals can also cut through skin, exposing a worker's blood to unwanted contamination.

Avoiding Chemical Exposure

- Use PPE as required.

- Never eat, drink, or smoke while using hazardous chemicals.

- Always read the chemical's SDS prior to use.

- Make sure all chemical containers are properly labeled.

- Always wash up after using chemicals.

- Never smell or taste a chemical to identify it.

- Know and practice all emergency evacuation and containment procedures and equipment.

- Store all hazardous chemicals properly.

- Always use hazardous chemicals as intended.

- Avoid creating aerosols in the laboratory: do not use open vessels for processing chemicals.

- Use chemical fume hoods when working with hazardous chemicals.

For Laboratory Safety Guidance, go to **www.osha.gov/sites/default/files/ publications/OSHA3404laboratory-safety-guidance.pdf**

For more information, go to the US Chemical Safety and Hazard Investigation Board's website: **www.csb.gov.**

Listen Up to Protect Your Hearing

A good analogy to explain how hearing loss occurs is to visualize a thick grassy lawn. As you walk across the grass, the grass bends down because of your weight. After you pass, the grass stands back up. The more you walk across the same area, the longer it takes for the grass to stand back upright. If you continue to walk across the same area, eventually the grass will die, and the area becomes a dirt path.

The same thing can happen to your hearing. When sound vibrations enter your ear, tiny hair cells in the inner ear change the vibrations into nerve impulses. The nerve impulses are then transmitted to the brain where they are translated into the sound we hear. When the hair cells are subjected to excessive noise, they begin to lie down just like grass does when we step on it. After the noise subsides, the hair cells stand back up. Over time, the more noise the hair cells are exposed to, the longer it takes for them to stand back up. Eventually, they fail to return to normal, resulting in permanent hearing damage.

Prolonged exposure to noise levels above 85 decibels (dB) can be harmful to hearing. The higher the decibel level of noise you are exposed to, the shorter the time you are allowed to work around the noise. CDC says that regular exposure to 110 dB for more than 1-minute risks permanent hearing loss. This is the level of sound an average chainsaw makes. An ambulance siren is about 120 dB.

When the noise levels vary, a mathematical calculation is used to determine a time-weighted average of the noise exposure (11 dB = 0.5 hour). If the sound level is a constant 95 dB, you would be able to work in a noisy environment for a total of 4 hours out of an 8-hour work shift. If the sound level was a constant 100 dB, you would be able to work a total of 2 hours.

Wear the Right Ear Gear

Courtesy of OSHA

Noise exposure can be reduced by wearing properly fitted hearing protection. All hearing protection must be labeled to show its effectiveness. The effectiveness is rated via the noise reduction rating (NRR). The higher the NRR, the more protection provided. Additional protection can be obtained by wearing an earmuff over earplugs. Don't be fooled, however, into believing that the protection will be the total of both NRRs added together; the increased protection will only muffle about 2 to 5 dB.

The highest NRR is typically provided by moldable earplugs—if they are worn correctly. They can be made of foam, wax, silicone, or other materials and fit directly in the ear canal. The earmuff, which can be custom fitted, has the next highest NRR. The least effective are semi-insert plugs—two earplugs held over the ends of the ear canal by a rigid headband. But remember, there can be a wide range of NRRs for the same type of protection. Read the label and follow the manufacturer's recommendations for wearing and maintaining the products.

For more information, go to OSHA's Hearing Protection Program website at **www.osha.gov/noise/hearing-programs**

Climb Onto Ladder Safety

Year after year, falls from ladders rank as one of the leading single causes of occupational fatalities and injuries. Fall protection and prevention are ongoing major concerns of OSHA. When possible, it is best to avoided using portable ladders. New facilities can be designed to reduce the need for ladders and, when possible, personnel lifts should be used instead.

If using a portable ladder is unavoidable, the following are safety tips to keep in mind.

Inspection:
- Before you use the ladder, inspect it for cracked or broken parts such as rungs, steps, side rails, feet, and locking components. By law, if it has any damage, it must be removed from service and tagged until repaired or discarded.

Positioning:
- Ladders to be set on stable, level ground to keep it from slipping or moving. You can lose your balance by simply getting on or off an unsteady ladder. Do not try to make the ladder reach farther by setting on block, boxes or other unstable bases.

- Position the ladder so its side rails extend at least 3 ft above the landing. When a 3-ft extension is not possible, secure the side rails at the top to a rigid support and use a grab device.

- To provide the correct angle so extension ladders won't slip, place the base of the ladder one foot away from whatever the top of the ladder leans against, for every four feet in height of the ladder. To check, put your feet at the base of the ladder and extend your arm straight out. If you can touch the closest part of the ladder without bending your arm, or bending over, the ladder is at the correct angle. If not, the ladder is not at a safe angle.

- Avoid electrical hazards. Never use a metal ladder near power lines or exposed energized electrical equipment. Look for overhead power lines before raising the ladder, and never allow the ladder to get closer than 10 ft to power lines. Also make sure that once you've climbed the ladder, your body and tools cannot come in contact with the power lines.

- A ladder placed in any location where it can be hit or displaced by other work activities must be secured, or a barricade must be erected to keep traffic away from the ladder.

- Be sure all locks on an extension ladder are properly engaged.

Working:

- Never put more weight on the ladder than it is designed to support. And be sure to include the weight of the tools and materials you are using. The safe weight load should be labeled on the ladder.

- Never use a self-supporting ladder (such as a stepladder) as a single ladder or in a partially closed position.

- Never use the top step/rung of a ladder as a step/rung unless it was designed for that purpose.

- Always maintain a three-point (two hands and a foot, or two feet and a hand) contact on the ladder when climbing.

- Keep your body near the middle of the step and face the ladder while climbing.

- Only use ladders and appropriate accessories for their designed purposes.

- Keep the rungs free of wet or slippery materials.

- Do not try to move or shift a ladder while a person or equipment is on the ladder.

For additional safety information, visit the American Ladder Institute: **www.americanladderinstitute.org/page/BasicLadderSafety**

Additional guidance from OSHA is found at **www.osha.gov/sites/default/files/publications/portable_ladder_qc.pdf**

Working at Height:
Don't Fall Into Danger

- *A 39-year-old worker died after falling 40 ft when a scaffolding suspension rope broke. He was a member of a three-man crew engaged in the abrasive blasting and painting of the interior of a 48-ft-high, 30-ft-diameter steel water tank. At the time of the accident, the victim was standing on an outer end of the scaffold platform and was pulling on the suspension rope to raise that end of the scaffold. He fell when the rope broke and his end of the platform dropped to a vertical position. The victim was not using personal fall protection equipment, although it was available and was being used by a second painter. An investigation revealed that the 5/8-in. hoist rope broke at a point where it had been burned some time before the incident.*

- *A worker was killed, another was injured, and a coworker was left hanging when a scaffold rope broke while they were painting the side of a building in San Francisco. They were on a two-point suspension scaffold that did not have guardrails; the ropes suspending the scaffold were old and had not been inspected; and the employees were not wearing safety belts. When the left scaffold rope broke and the scaffold collapsed, one painter was killed while another fell to a nearby roof and broke both arms. The coworker was left hanging on to the remaining scaffold rope.*

These case reports from OSHA files demonstrate the dangers of a failure to have a good fall-protection, fall-arrest, and fall-restraint system in place for employees. Anyone working at height needs to be aware of the dangers and have the proper protection in place to prevent such accidents.

Falls can occur from floor and roof edges, elevated platforms, ledges, elevated tanks, machine rooms, and attached ladders and stairways. Falls can also occur from temporary structures used for construction and maintenance such as scaffolds or ladders.

OSHA standards state that a guardrail system, safety net system, or personal fall-arrest system must be in place to protect workers who are exposed to an unprotected side or edge from which they could fall 6 ft or more.

Built-in Safety

The safest strategy for workers who must work at elevation is to have built-in fall restraints, such as

- permanent guardrails that meet OSHA height and strength requirements,
- built-in anchor points with appropriate personal fall-arrest systems and lifelines, or
- other forms of fall protection such as safety netting.

On-the-Job Safety

Workers at elevations with vertical drops of 6 ft or more should also be provided a personal fall-restraint system that secures the worker via an anchor point, connector, lanyard, and body harness. This system is designed to prevent a fall.

A personal fall-arrest system also uses an anchor point, connectors, lanyards, and body harnesses but allows exposure to the fall and is then designed to stop the fall after it has begun.

Key points about a personal fall-arrest system include the following.

- Connectors should be made of drop-forged, pressed, or formed steel or of equivalent materials and covered with a corrosion-resistant finish, with smooth surfaces and edges to prevent damage to interfacing parts of the system.
- D-rings and snap hooks should have a minimum tensile strength of 5,000 lb and be proof tested to a minimum tensile load of 3,600 lb without cracking, breaking, or becoming deformed.
- Locking snap hooks must prevent disengagement of the snap hook if the connected member contacts the snap hook keeper.
- Unless designed for it, locking snap hooks must not be attached
  - directly to webbing, rope, or wire rope;
  - to each other;
  - to a D-ring to which another snap hook or other connector is attached;
  - to a horizontal lifeline; or
  - to any object that is incompatible with the snap hook such that unintentional disengagement could occur.
- Horizontal lifelines should be designed, installed, and used (under the supervision of a qualified person) as part of a complete personal fall-arrest system that maintains a safety factor of at least two.
- Lanyards and vertical lifelines should have a minimum breaking strength of 5,000 lb.

- When vertical lifelines are used, each person must be attached to a separate lifeline.

- Lifelines must be protected against being cut or abraded.

- Self-retracting lifelines and lanyards that automatically limit free-fall distance to 2 ft or less must be able to sustain a tensile load of at least 3,000 lb.

When stopping a fall, personal fall-arrest systems should

- limit the maximum arresting force on a person with a body harness to 1,800 lb,

- prevent a free fall of more than 6 ft or contact with any lower level,

- bring a person to a complete stop and limit maximum deceleration distance to 3.5 ft, and

- withstand twice the potential impact energy of a person free falling a distance of 6 ft, or the free fall distance permitted by the system (whichever is less).

Use body harnesses and components only for personal protection, and never use harnesses to hoist materials. Inspect personal systems before use for wear and damage. Any fall-arrest systems and components that are subjected to impact loading must immediately be removed from service and not used again for protection until inspected by a competent per-son and determined to be undamaged and suitable for reuse.

For more information, see the OSHA website: **www.osha.gov/fall-protection**

The Safe Use of Compressed Air

When compressed air is commonly used by water operators to perform countless jobs from filling tires, lubricating trucks, and operating lifts, to breaking, jacking, auguring, and tamping of earth and rock on construction projects. Compressed air helps us complete our jobs better and faster. This discussion is geared to its use and the dangers of its misuse.

The misuse of compressed-air hose is hazardous, especially if the air stream is brought into close or direct contact with any portion of a worker's body or clothing. If this occurs and the skin is broken, air may be forced into the bloodstream, often with fatal results.

Air in the bloodstream is just one-way compressed air can injure a worker. Using compressed air to clean clothing, tools, or workbenches can result in foreign bodies in workers' eyes. Using compressed air for cleaning is not only an unsafe practice but more of a hindrance than help because it spreads dust and chips around, which eventually results in a larger cleanup area.

Unfastened safety chains on air-hose lines account for more injuries than any other type of compressed-air accident. Hose couplings can be handled roughly on construction jobs—e.g., dragged over the ground or streets—which can lead to the disconnection of couplings. That's why a safety chain must be connected from one hose to the other at each connection. Once an unchained hose is accidentally disconnected, escaping high-pressure air can whip the hose around with terrific force, causing the hose to strike anything in its path.

Safety Tips for Using Compressed Air

- All operators should be trained on how to use equipment safely and avoid injury.

- Personal protective equipment such as safety glasses and hearing protection should be worn whenever you use compressed air.

- Compressed air shall not be used for cleaning purposes except where reduced to less than 30 psi.

- Carefully inspect all equipment to see that it is in good shape.

- Check that pipes, hoses and fitting have the same pressure rating as the compressor.

- Make sure the shut-off valve is within reach of the operator.

- Check the run of air hose to see that it is protected from possible damage and is not a tripping hazard.

- Be sure the valve is closed on the supply side of the coupling before changing the tool at the end of a compressed air line. No matter where the valve is—close it. Never simply kink the hose.

- After closing the valve, pull the trigger or open the operating valve to release the line pressure. Then make the required tool change.

Practice these safety measures when working with compressed air. For more information go to the OSHA website: **www.osha.gov/compressed-gas-equipment**

Temperature Extremes:
Hypothermia and Heat Exhaustion

Fresh air and sunshine can be benefits of working outdoors, except when it gets uncomfortably cold or hot. However, temperature extremes are much more than a matter of comfort. They can also cause health hazards with deadly consequences.

It's important that you and your coworkers know how to recognize the symptoms of hypothermia, frostbite, and heat-related illnesses and how to respond to the effects.

Hypothermia

Hypothermia is a life-threatening condition that occurs when the body core loses heat faster than it can be generated. Obviously, hypothermia can occur in the winter, when the weather is cold, but it can happen during any season; for example, in the summer, when someone is immersed in water that is colder than body temperature for an extended period of time or working in a cold meter pit underground for a long time.

The early symptoms include uncontrollable shivering, impaired or slurred speech, and awkward or clumsy body movements.

As the body temperature continues to drop, nausea, apathy, confusion, and lethargy can also occur. Often a severely affected victim will lie down, fall asleep, or lose consciousness. The final stages can result in coma and death.

If you identify any of the above symptoms in yourself or someone you are working with, take the following steps immediately:

- Get the victim to a warm location that is sheltered from the wind.

- Remove all wet clothing and anything that might restrict circulation. Cover the victim's body and head with warm, dry clothing or blankets.

Rewarming should be started by applying warm compresses to the chest, neck, and groin. If necessary, body-to-body contact can be used as a first-aid measure. This passive rewarming approach may be all that is required for a conscious person who is shivering. Hot water and direct heat should never be applied!

If the victim does not respond and the symptoms become progressively worse,

- call 911 immediately in accordance with your emergency plans,

- monitor the victim's breathing and start CPR if the breathing seems dangerously slow or stops, and

- keep the victim immobile until medical help arrives.

Winter: Frostbite

Frostbite occurs when the fluids and underlying soft tissue of the skin freeze. It is typically accelerated by wind and humidity. That is, although the temperature is above 32°F (0°C), the skin may still freeze because of a wind chill factor. The most commonly affected areas are the nose, cheeks, ears, fingers, and toes.

Symptoms of frostbite include gray or yellowish patches of skin. The affected parts are usually numb but feel cold. Pain is sometimes felt early but later disappears. The skin remains soft and flexible, but after it thaws, it becomes red and flaky.

If the frostbite is severe (deep), the skin is generally waxy and pale and may turn blue or purple when thawed. Large blisters may also appear.

First aid for frostbite includes bringing the victim indoors and providing warm non-alcoholic beverages. Warm the frozen area by immersing it in warm water (not hot!) or wrapping it in blankets or clean clothing. Do not rub the affected area; that can lead to gangrene. Obtain medical assistance as soon as possible.

Summer: Heat-Related Illnesses

When your body heats up faster than it can cool itself, mild to severe illnesses may develop. Air temperature, humidity, and clothing can increase the risk of developing heat illnesses. Age, gender, weight, physical fitness, nutrition, alcohol or drug use, or pre-existing diseases, like diabetes, can also increase the risk. Heat-related illnesses include

- heat rash (prickly heat)—when the sweat ducts to the skin become blocked or swell, causing discomfort and itching;

- heat cramps—when muscles cramp up after exercise because sweating causes the body to lose water, salt, and minerals (electrolytes);

- heat edema—when legs and hands swell after sitting or standing for a long time in a hot environment;

- heat tetany (hyperventilation and heat stress)—usually caused by short periods of stress in a hot environment;

- heat syncope (fainting)—when a person suddenly loses consciousness because of low blood pressure from heat, which causes the blood vessels to dilate, and gravity moves body fluids into the legs;

- heat exhaustion (heat prostration)—usually happens when a person is working or exercising in hot weather and does not drink enough liquids to replace those lost liquids; and

- heat stroke (sun stroke)—when the body fails to regulate its own temperature and body temperature continues to rise, often to 105°F (40.6°C) or higher. Heat stroke is a medical emergency. Even with immediate treatment, it can be life-threatening or cause serious long-term problems.

Knowing how to recognize the early symptoms of heat illnesses and knowing how to prevent, control, and respond to the effects can help make everyone's job safer.

Preventing or Controlling Heat Illnesses

- Drink about a cup of cool water every 15–20 minutes. Avoid caffeine, sugary drinks, and alcohol. Use sports drinks in moderation.

- Limit exposure time to the heat; schedule hot jobs for cooler times of the day. Take frequent rest breaks in cool areas.

- Gradually adapt yourself to the heat. It takes up to 10 days for your body to adapt to high heat.

- Slow your pace and try to mechanize heavy jobs.

- Wear loose, lightweight clothing and a hat, and protect exposed skin.

- Do not use salt tablets.

If skin rash, stomach cramps, fatigue, or dizziness occur, the victim needs to immediately seek rest in a cool, shady place, drink lots of water, and repeatedly wet and dry the skin.

If the symptoms increase to excessive sweating; cold, moist, pale, or flushed skin; thirst; extreme fatigue; headache; nausea; or a rapid pulse, the victim may be experiencing heat exhaustion. The victim should immediately lie down in a cool, shaded place and sip lots of cool water until the symptoms disappear. If the symptoms worsen or the victim becomes unconscious, immediately get medical help according to your utility's emergency procedures.

Severe heat illness can lead to heat stroke, which can be fatal or lead to permanent brain damage if the victim does not receive immediate medical treatment. Unfortunately, there's little warning when a victim reaches this crisis stage.

If a victim's skin becomes hot, dry, red, or spotted and the victim experiences confusion, delirium, or convulsions or slips into unconsciousness, the person is likely experiencing heat stroke and urgently needs medical help. While waiting for that help to arrive, loosen the victim's clothing and pour water over the entire body. Never try to force an unconscious victim to drink water.

For more information, go to the WebMD website on heat-related illnesses: **www.webmd.com/first-aid/tc/heat-related-illnesses-topic-overview** or the national public service site "Ready" for information on both topics: **www.ready.gov/ winter-weather**; **www.Ready.gov/heat**. For additional information on extreme weather safety, please go to these OSHA websites: **www.osha.gov/heat-exposure** and **www.osha.gov/winter-weather.**

Cutting Pipe Safely With Power Saws

Gas, hydraulic, and pneumatic saws are all used to cut utility water pipes. The main difference is the type of blade used. Depending on the situation and type of pipe, a specific blade may be required; some blades, such as diamond blades, will cut a variety of materials. So be sure to choose the proper saw and blade for the material, following the manufacturer's recommendations for the type of finish that is needed. Using a saw or blade not designed to efficiently cut through a material will usually damage the tool and create a safety hazard for its operator. Forcing a saw that is not big enough for the job can cause a kickback.

Users should be provided with specific tool training and have read the entire operating manual and manufacturer's guide for the specific saw used on the job.

Before Each Use
Carefully examine the cutting equipment. Look for the following:

- worn bearings,
- damaged power cords,
- faulty on/off switches,
- loose bolts or nuts,
- lubricant leakage,
- evidence of excessive rust and broken or damaged housing or casing.
- Inspect the cutting blade or chain to ensure that it is sharp;
- is not crooked, bent, cracked, or split;
- rotates in the proper direction; and
- is securely fastened or bolted into place and does not wiggle loosely if you try to tap or vibrate it gently by hand.

- In addition, check the safety guards to make sure they are in place and secure and that the machine warning placards and labels are in place and legible. Follow the manufacturer's recommendations on blade replacement and preventive maintenance.

If the saw is damaged or needs servicing, put a tag on it indicating it should not be used. Mark on the tag what is wrong with the unit, and arrange to either have the unit repaired, serviced, or disposed of.

Before Cutting

- Wear the appropriate PPE, including head protection with safety glasses and/or a face shield; hearing protection; respiratory protection if necessary; steel-toed safety shoes; close-fitting clothing and long trousers or, for chainsaws, special ballistic nylon reinforced chaps, pants, gloves, and boots.

- Properly support and chock the pipe to be cut so it won't move or flex during the cut.

- Fuel the saw, as appropriate, with the proper oil–gas mixture. Never gas, lubricate, or service a running machine.

- Clear the immediate area of people, tools, debris, and other obstacles.

While Cutting

- Maintain good footing, with your feet shoulder-width apart.

- Keep the saw close to your body. Don't reach with the saw.

- Position your body as close to the pipe as possible. Don't reach with the cutting tool.

- Work with slow, controlled movements. Bend your knees if necessary.

- Don't rock the saw back and forth; allow the weight of the saw to help pace the cut. Never twist or turn the blade while cutting; make a straight, even cut.

- Work at a steady pace; never force the blade through the material.

- Stay focused on the task of cutting.

Other Safety Measures

- Always allow the saw to turn off, power down, and stop moving before you take your attention away from it. Never leave the machine unattended while running.

- Keep in mind that cutting and grinding is considered hot work. Never cut in the vicinity of flammable materials or in areas without proper ventilation.

- Have first-aid kits, fire extinguishers, and emergency call numbers in close proximity at all times.

- Don't exceed the maximum operating speed recommended by the manufacturer, and never cut material not listed by the manufacturer of that saw and blade.

- The use of gas-powered pipe saws within excavations requires the use of ventilation equipment to prevent carbon monoxide accumulation.

- If a blade or saw appears to overheat, turn the tool off immediately and allow it to cool down. After it has cooled, check it carefully to make sure neither the saw nor blade has been damaged.

- If you are cutting pipe that is concrete or asbestos, ensure you take the proper precautions to prevent dust or use respiratory protection to prevent inhalation of silica dust or asbestos fibers.

For more information, visit **www.osha.gov/silica-crystalline**

Vehicle Safety:
Check, Inspect, Drive!

Using a company vehicle means you have a responsibility to ensure not only your own safety but also that of your passengers and fellow drivers. Because the vehicle likely has been driven by other people, it's a good idea to take a few minutes before you drive to check that the vehicle and its equipment are in safe and proper working order.

Vehicle Safety Checklist					
Vehicle Number	Item	Good	Needs Attention	Employee	Inspection Date
	Lights (including emergency flashers)				
	Horn				
	Mirrors & Visors				
	Windshield (including wiper blades & washer fluid)				
	All Glass				
	Brakes & Parking Brakes				
	Tires & Wheels				
	Seat Belt & Shoulder Harness				
	Interior Condition (floor mats, seats, dashboard)				
	Exterior Condition (including locks)				
	State Inspection & State License				
	County, City, or Town License & Safety Stickers				

Vehicle Safety Checklist					
Vehicle Number	Item	Good	Needs Attention	Employee	Inspection Date
	First-Aid Kit & Accident Report Kit (includes insurance card)				
	Ladders & Other Equipment				
	Exhaust System (muffler & tailpipe)				
	Fire Extinguisher				
	Logos & Vehicle Numbers				
	Tow or Trailer Hook				
	Items Secured in Vehicle				
	Additional Items				

Safe Fuel Handling Practices

The safe handling of gasoline and diesel fuels is everyone's responsibility. You can take steps to ensure that your own safety and health, as well as that of those around you and the environment, are protected. The improper handling of fuel can result in serious injury or death caused by fire, explosion, or asphyxiation.

Environmental Safety

Fuel released into the environment contaminates soil and groundwater. As a water utility worker, you know that contaminated groundwater supplies can sicken people and animals. Gasoline vapors are also harmful to human health—even at low concentrations—and are especially dangerous at high concentrations.

Here are some safety tips for what you can, and should, do to ensure safe fuel handling.

Safe Fueling

- Turn off the engine before fueling.

- Never smoke or light matches or lighters while fueling.

- Stand upwind of the nozzle while refueling and try to not breathe the fumes.

- Do not top off the tank. Even the little drips that fall onto the pavement can contaminate soil, groundwater, or surface water.

- Do not leave your vehicle unattended while the pump is running.

Use the Proper Container

- Use only containers approved by a reputable testing lab, such as Underwriters Laboratories.

- Keep the container tightly sealed.

- Containers should be fitted with a spout to allow pouring without spilling and to minimize the generation of vapors.

- Keep gas containers out of direct sunlight.
- Always open and use gasoline containers in a well-ventilated area.

Safe Storage

Gasoline moves quickly through soil and into groundwater. Therefore, store and use gasoline and fuel equipment as far away from water wells as possible.

- Fill cautiously, and store no more than 10 gallons.
- Keep a closed cap on the gasoline container.
- Store the gasoline in a cool, dry place.
- Store at ground level, not on a shelf. Ground-level storage minimizes the danger of the container falling and spilling.
- Do not store gasoline in a vehicle's trunk, where it could explode from heat or impact.
- Always use a funnel and/or spout to prevent spilling or splashing when fueling portable and mobile equipment.
- Always fuel outdoors where there is good ventilation to disperse the vapors.
- Fuel equipment on a hard surface such as concrete or asphalt, rather than on soil or water.
- Portable cans and fuel tanks should be removed from the vehicle and filled while on the ground. A secondary containment device under the tank ensures even better spill protection.

Avoid Spills

Spilled motor fuels impact the environment through evaporation into the air, diffusion into the soil, and release into groundwater. Each year, Americans spill more than 9 million gallons of gasoline—the equivalent of an oil supertanker. The environmental impacts of improper handling, storage, and disposal of gasoline largely stem from sloppy filling of small engines, using inappropriate containers, overfilling motor equipment engines, storing gasoline in open containers, and disposing of excess gasoline improperly. If a spill occurs, use kitty litter, sawdust, or an absorbent towel to soak up the spill, then dispose of it properly.

Safe Disposal

Do not dispose of gasoline down the drain, into surface water, onto the ground, or in the trash. Use the local hazardous waste collection and disposal location for safe and convenient disposal of excess or old gasoline.

For more information go to the US Environmental Protection Agency website on gasoline standards and programs: **www.epa.gov/otaq/fuels/gasolinefuels**

Keeping Chemical Deliveries Safe

One of the most potentially dangerous activities at a water or wastewater plant is the delivery of hazardous chemicals. Injuries can occur when a chemical is delivered to the wrong container or the chemical is spilled. Serious injury can result when incompatible toxic chemicals are inadvertently mixed during delivery.

Safe chemical receiving and unloading procedures, practices, and management controls should be documented and practiced to ensure the safe delivery of chemicals to utility facilities.

While the supplier and shipper are responsible for ensuring the chemical load is properly identified, placarded, and transported, it is up to the staff of the receiving facility to ensure that

- the chemical is what was ordered,
- the chemical is offloaded safely to the proper place, and
- personnel are trained to handle the material correctly.

In addition, facility security protocols should be used to verify in advance who will be delivering the chemicals and in what manner. A chain of custody should always be maintained between the manufacturer and the purchaser. It is even common for some utilities to require background checks on the chemical delivery drivers for their specific utility.

While every chemical has specific safety precautions that must be taken when inspecting and handling (and these are spelled out in the SDS), the following general procedures should be followed during the delivery and acceptance of any chemical.

1. Schedule the delivery so the proper personnel are onsite when the chemical is delivered and the facility is ready to receive the delivery.

2. Confirm the identity of the delivery driver; verify he or she is who the supplier had scheduled to make the delivery and that the vehicle or cargo container is the same transport container that is listed on the manifest. Verify the contents of the container. Read the placard, the bill of lading, and the SDS, and do any testing necessary. AWWA chemical standards state, "Each product

shall be identified as to product, grade, net weight, name and address of the manufacturer, and brand name. Packages or containers shall show a lot number and identification of manufacturer. All markings on packaged, containerized, or bulk shipments shall conform to applicable laws and regulations, including requirements established by OSHA. Bulk quantities of product should be sealed with a uniquely numbered tamper-evident seal(s)."

3. Wear any and all required PPE.

4. Inspect transferring hoses, valves, and recipient containers for damage, plugging, or wear, and replace as necessary.

5. Have a trained attendant oversee the unloading of cargo tanks. This person must be alert and

   - have a clear view of the cargo tank,

   - be within 25 ft of the tank,

   - be aware of the hazards,

   - know the procedures to follow in an emergency, and

   - be authorized to move the cargo tank and be able to do so if necessary.

6. Inspect the actual container or pipe that the product should be loaded into and/or through and be sure the receptacle is clear of all potential contaminants.

7. Unhook all loading/unloading connections before coupling, uncoupling, or moving a chemical cargo tank. Always chock trailers and semitrailers to prevent motion after the trailers are dropped.

8. Unless the engine must run a pump for product transfer, turn it off when loading or unloading.

9. Ground tanks correctly before filling them with flammable materials through an open filling hole. Ground the tank before opening the filling hole and maintain the ground until the filling hole is closed.

10. Close all manholes and valves before moving a tank carrying hazardous materials. It does not matter how small the amount in the tank or how short the distance, manholes and valves must be closed.

11. Keep liquid discharge valves on a compressed gas tank closed except when loading and unloading.

12. Ensure that any necessary lockout/tagout procedures are followed.

13. Know what to do in the event of a spill, chemical release, or individual exposure to a chemical.

For more information, go to the American Chemistry Council's website: **www.americanchemistry.com/Safety/TransportationSafety**, or the US Chemical Safety and Hazard Investigation Board's website: **www.csb.gov**

Don't Get in a Bind With a Backhoe

The backhoe is a highly productive machine—the true workhorse for most projects involving trenching and eatrth moving. But a backhoe is also a complicated and dangerous machine that requires continuous vigilance during its operation. Backhoe operators are responsible for analyzing and reacting to all situations to keep fellow workers safe from potential accidents.

The best way to operate a backhoe safely and efficiently is to understand the jobsite, the equipment, and, as a driver, yourself.

Before Starting Work

- Always review the manufacturers operator manual prior to use to understand all the features, pre-use inspection, and safe operating procedures.

- Make sure the machine is fit for the task. Walk around the machine and inspect it with care. Look for damaged or missing parts, and check for fluid leaks, cracks, and excessive wear. Make sure the control levers are working properly.

- Select the right size bucket for the job. Make sure it matches the workload.

- Only use implements or attachments designed or rated for the backhoe you are operating.

- Review the equipment's warning and safety signs. They are there for a reason. Take the signs seriously and heed their warnings. Replace any damaged or missing decals.

- Inspect the jobsite. Is it safe for the backhoe? Stake out the area to be excavated using marker flags. However, do not disturb the markings made by the underground utility locating service.

- Be sure to always look up for overhead power lines. If power lines are on the site always keep them firmly in mind and point them out to your coworkers. Never allow a fully extended boom to get any closer than 10 ft from a power line—greater than 10 ft is even better! And never move the machine while the boom is elevated. Never work in areas that have inadequate overhead clearances. It is just too dangerous.

- Call before you dig! Did you call 811 two working days in advance so the locations of all underground utilities, in addition to water, are clearly marked at the construction site? Don't rely solely on your company's charts. You need to be certain.

- Be honest and ask yourself: Am I qualified to operate the equipment? To be a qualified backhoe operator, you should not only have mastery of the operating skills but also have a strong sense of safety. Good operators will instinctively focus more on their safety sense than on their operating skills.

Backhoe Basics

A backhoe operator needs to know how to operate both a front-end loader and a backhoe scooper. The front-end loader is not as complicated as the backhoe attachment, but the operator must use a joystick control while simultaneously driving the tractor. The front-end loader will either remove excess dirt and material from the site or place it back in the trench. The front-mounted bucket can also tamp down loose soil and create a level grade.

Even though backhoe models vary, all have a few standard safety features. These include steps and grab handles for getting on and off of the machine. Frame lock levers and attaching levers keep the backhoe securely fastened to the loader frame during operation, as well as when it's being transported.

Some backhoes provide a safety chain to prevent the backhoe mounting frame from rotating backward and unexpectedly trapping the operator. Therefore, it is important to know and check all the mounting and attachment points and the safety chain before operating the backhoe.

Check the loader/backhoe to be sure the following safety devices are in good working order.

- Rollover protective structure (ROPS)
- Seat belt (if ROPS equipped)
- Guards
- Shields
- Backup warning system
- Lights and mirrors
- Fire extinguisher

The Right PPE

Wearing the appropriate PPE is important when operating a backhoe. In addition to sturdy pants and shirt, safety shoes, gloves, and a hard hat, the work may require

- safety goggles or glasses,

- hearing protection for prolonged exposure,

- a respirator for dusty conditions, and

- High Visibility clothing if working near high traffic areas or roadways.

- Review safety data sheets (SDS) for fuels, oils and fluids required for the operations and maintenance of the backhoe for any additional safe handling instructions.

Operating the Backhoe

- Operate the backhoe only from the seat.

- Always lower the stabilizer feet to provide extra grip and leverage. Level the machine for maximum stability.

- Never allow riders or passengers to ride on the backhoe while being operated.

- Keep bystanders and other workers out of the bucket swing area. Always be aware if other people are around you and where they are standing.

- Make sure there's enough clearance to swing the loader bucket to one side for dumping. Keep the bucket low to the ground.

- Double-check the lock on the backhoe attachment.

- Never swing the bucket over a truck cab.

- Dump the bucket uphill if possible when operating on a slope. If you must dump downhill, swing slowly to avoid tipping the machine.

- If using the backhoe as a hoist, do so with the weight over the back of the machine—never to the side—to avoid tipping.

- Be sure the load you are lifting is balanced and move the boom slowly to avoid swaying the load.

- About every 8 hours, grease all of the Zerk fittings. Check the hydraulic fluid and oil daily. If the fluid is low, the backhoe will not operate properly.

- A clean backhoe is a safe backhoe. Keep engine compartment, radiator, batteries, hydraulic lines, fuel tank and operator's station clean.

- Any time you leave the operator seat or lower the bucket or attachment to the ground, turn the engine off and remove the ignition key.

Transporting the backhoe

- Follow DOT requirements, local rules, and manufacturers procedures when transporting the backhoe and other heavy equipment to and from work sites. Ensure backhoe is properly secured during transport. If driving on or operating near public roadways, a slow-moving vehicle (SMV) sign will be required.

Many heavy equipment manufacturers have free backhoe safety videos on YouTube and safety tips on their websites.

Operating Laboratory Equipment

The cleaner, safer, and more efficient a lab is, the better and more reliable the data coming out of it will be. It is easy to make safety a priority in a water utility's laboratory by learning the proper safety procedures for the lab's equipment.

Water utility labs may be similar, but the equipment in them can vary or have different operating instructions. Lab technicians should become familiar with all the equipment in each lab and be trained on any new equipment brought into it.

Here are some basic procedures for equipment one could expect to find in most water laboratories.

Fume Hood

- Fume hoods are used to prevent toxic and flammable vapors from entering the general lab area. They provide a protective barrier between lab personnel and where chemical reactions are taking place.

- Keep chemicals and substances in a fume hood to a minimum when possible and be sure that nothing is blocking fume hood vents. Proper airflow into a lab hood should be verified on a periodic basis.

- When working with a hood partially open, make sure materials are pushed back away from the front of the hood area to keep vapors inside the hood.

- For some procedures, it is necessary to keep the hood securely closed.

Several pieces of lab equipment are used to heat substances: autoclaves, controlled temperature hot plates, steam-heated ovens, and Bunsen burners.

Bunsen Burner

- Bunsen burners are obviously an ignition source, so be aware of a chemical's flashpoint and explosive limits.

- Place the burner in a position where it is unlikely a worker will reach across the flame during any lab procedure.

- When using autoclaves, ovens, and hot plates, be sure to use tongs, gloves, and other accessories to prevent burns.

Autoclaves

- Autoclaves are pressure vessels that use high-pressure steam to sterilize lab equipment to kill micro-organisms including bacteria, fungi, and viruses.

- Always follow manufacturers operating procedures for proper use, inspection and maintenance of autoclaves.

- Never attempt to open an autoclave while under pressure.

- Wearing appropriate Personal Protective Equipment (PPE) should include a lab coat, heat resistant gloves rated for the temperatures exposed to when using an autoclave, and eye protection when unloading the autoclave.

Electrical Cords

- Any piece of lab equipment that is powered by electricity has a risk of static electricity buildup. Make sure all appliances are properly grounded and do not overload laboratory circuits.

- Cords to lab equipment should have acid and waterproof insulation.

- Cords should be inspected often to make sure they have not become frayed or loose at the connections and should be replaced when necessary.

- Electrical equipment involving wet processes or use in areas within six feet of a water source shall be used with a ground-fault circuit interrupter (GFCI) outlet and tested regularly.

Lab Equipment Safety Guards

- Make sure safety guards installed in any moving machine in the lab are in place and functioning correctly.

- The lab should have a checklist that indicates to workers what pieces of equipment should be left on overnight and what equipment needs to be shut off.

Tanks of Compressed Gases

- Compressed gases should be stored based on local fire codes and manufacturer recommendations.

- Store gas tanks upright with a restraining chain.

- Store gas tanks out of direct sunlight and away from excessive heat sources.

Lab Glassware

- Never use cracked or chipped glassware.

- If glassware is broken, immediately clean it up while wearing protective gloves and put it in a container designed to hold it.

- When carrying a large glass container, use both hands, with one supporting it from the bottom.

Test Tubes

- When heating chemicals in test tubes, remember that the chemicals can be super-heated and fly out or boil over—so always point them away from yourself and others.

- Use a towel or wear gloves when connecting a rubber stopper to glass tubing. Use a compatible lubricant to help make the connection secure and be sure to clamp it.

- Always use suction bulbs with pipettes; never pipette with your mouth.

Hazardous Procedures

- Never conduct hazardous procedures alone. If working in a small lab with limited personnel, rely on other members of the utility to observe these procedures.

- Some lab procedures continue while technicians are gone. For unattended procedures, always leave the lights on and a sign on the door warning others about the lab procedures in progress.

- Know the location of all safety equipment, nearest fire pull-station, emergency number, and first aid treatment kit.

After Working in the Lab

- When you are done working in a lab, remove your lab coat before leaving the lab to avoid contaminating other areas of the utility.

- Wash all possible areas of body exposure with soap, in a shower if available.

- In fact, it's a good idea to wash up every time you enter and exit a lab and follow each lab procedure.

- Clean lab areas you are working in after each procedure and at the end of the day.

- Empty the trash and waste material often.

- Remove tripping hazards and keep the lab floor clean and dry.

- Clean up even the smallest spills immediately.

- Water utility janitorial staff should be trained in lab cleanup procedures as labs have different requirements than other areas in the facility.

Considerations for Microbiology Labs

- Housekeeping is especially important here because microbials can contaminate a lab. Special precautions need to be taken to reduce the chances of exposure to the lab and technicians. Workers should be trained in the procedures and cleanup of microbial labs.

- OSHA requires employers to periodically measure lab workers' exposure to substances regulated by OSHA's laboratory standard. The results should be made available to workers and corrective action taken immediately by lab supervisors when necessary. The standard also requires the establishment of a Chemical Hygiene Plan, employee training and information, medical consultations and examinations, hazard identification, and recordkeeping.

Water utilities should also use engineering and administrative controls to reduce the risk of working in labs. Changing the work environment and modifying workers' schedules to reduce the exposure to workplace hazards are examples of strategies that can eliminate problems in advance. Also, encourage laboratory personnel to eliminate unsafe work practices by creating workplace safety rules and standard operating procedures.

Additional Resources

- For more information on lab safety, review OSHA's Laboratory Safety Guidance at **www.osha.gov/Publications/laboratory/OSHA3404laboratory-safety-guidance.pdf**

- National Research Council (US) Committee on Prudent Practices in the Laboratory.

- Washington (DC): National Academies Press (US); 2011.

Setting Up a Safe Traffic Control Zone

More than a thousand people are killed each year in work-zone traffic accidents. Eighty percent of those fatalities are drivers and their passengers. Speed and driver inattention are the leading causes of these preventable accidents. However, don't get too comfortable—according to the Occupational Safety and Health Administration (OSHA), employees in these highway work zones have one of the most dangerous occupations in the United States.

Following are a few simple tips for setting up a safe work zone.

- Any time working on or near roadways, you must have an established traffic control plan to control the safety of workers and drivers in a temporary traffic control zone.

- Expect the unexpected and never assume drivers see you.

- Understand that drivers may be confused, angry, or distracted when entering a work zone and may have difficulty negotiating the detours.

- When you set up a detour, try to avoid requiring drivers to make sudden lane changes or encounter unexpected road conditions.

- Always pay attention to the traffic. Beware of complacency.

- Never turn your back to oncoming traffic. If you do need to work with your back to the traffic, use a spotter. Have a communications plan between you and that spotter.

- When addressing traffic control, refer to state and national standards and statutes for traffic control devices and roadway worker safety.

- All roadside workers must wear bright and highly reflective ANSI Class 3 protective garments. These garments are recommended for both day and night use, and they meet the requirement to be visible from 1,000 ft at night.

- Flaggers need to stand on the shoulder and focus on approaching vehicles. Avoid standing in the lane unless visibility is an issue. Once traffic is stopped, flaggers should move back to the shoulder of the road.

Flagger Safety

Traffic flaggers actively manage the safe flow of vehicles, equipment, and pedestrians in temporary traffic control zones using hand-signaling devices or an Automated Flagger Assistance Device (AFAD). Their responsibilities are critical to the safety and welfare of their fellow workers, passing drivers, and pedestrians, therefore flaggers should be properly qualified and trained.

To be both safe and effective, flaggers need to understand the overall project, the flow of the construction work and the workers, the jobsite's equipment and machinery, and the ever-changing pattern of activities. They need to anticipate and adjust their work in fast-changing situations.

Two-way radio communication with the drivers of the construction equipment, with fellow flaggers with whom they need to coordinate traffic flow, and with the site manager is essential for maximum safety.

Perhaps the biggest mistakes a flagger can make are to get too comfortable with the job and to lose concentration.

Work-Zone Personal Protective Equipment

Head protection must be worn at all times. In all heavy construction areas, required foot protection includes steel-toe shoes with heavy-duty soles to help prevent crushing and penetration. Flaggers are on their feet most of the time, so their shoes need to fit well and be comfortable. Hearing protection includes earplugs or high-tech earmuffs.

For safety reasons, every worker should be able to hear the muted sounds of the construction site—and they should never wear headphones or headsets plugged into a music audio device. And don't forget a face mask for dust protection.

Frequent checks of the work-zone diversions and detours during construction will tell you if your temporary traffic control plan is being followed, that the traffic control devices are in their proper place and working, and that a safe, accessible pedestrian route is available at all times.

Holding on to Hand Safety

Every year, about one million US workers receive emergency hospital treatment for acute and serious hand, finger, and wrist injuries. Unfortunately, in one recent year, almost 8,000 of these injuries resulted in amputations.

According to OSHA, close to 70 percent of victims experiencing hand, finger, and wrist injuries were not wearing proper PPE. The other 30 percent wore gloves or PPE that were inadequate, damaged, or wrong for the type of work being performed. OSHA now requires employers to determine the most appropriate types of PPE for employees' hands based on the specific work conditions and potential workplace hazards of the task to be performed.

Many employers have found success in having their employees conduct their own hazard assessment for hand safety. It makes sense that involving employees in the assessment process increases their safety awareness. For example, when opening a discussion about hand safety, ask the employees to list all the ways their hands might be injured on a particular job. This list might include

- cuts, lacerations, punctures, and even amputations;
- abrasions from rough surfaces;
- broken fingers and bones in the hand;
- chemical burns and severe skin irritation;
- thermal burns from touching extremely hot objects; and
- absorption of hazardous substances through unprotected skin.

A study by the Liberty Mutual Research Institute for Safety found that wearing gloves reduced hand injuries by 60 percent. Although gloves will help protect against many of the above hazards, no glove protects against all hazards. You must select the right gloves for the hazards of the specific job.

The Right Glove

So, how do you select the right gloves for the job? As with any PPE selection process, the first step is to conduct a risk assessment to identify and understand the potential hazards.

Identify the substances (particulates, liquids, and gases) present in the worksite and the hazards associated with these substances. Survey the worksite and list all physical and environmental hazards such as sharp instruments, rough surfaces, or machinery. Also, make a list of employees who will be wearing the gloves, the work each person will do, and what equipment will be used. Keep in mind that some hand injuries (lacerations, crushing, broken bones, amputations) cannot be prevented by gloves.

Gloves should be evaluated by the following criteria:

- physical protection: resistance to cuts, punctures, and abrasions;
- chemical protection: resistance to permeation, degradation and penetration;
- full protection: no holes or tears;
- heat and flame protection;
- cold protection;
- vibration reduction;
- dexterity for the job at hand; and
- voltage rating.

In addition, consider other hand protection features such as length, size, coverage area, type of cuff, surface finish, and any attributes affecting function or comfort. Also consider the materials the gloves are made of. Select gloves that offer the optimal combination of features and performance. Periodically reevaluate your choices with your employees.

When it comes to the materials gloves are made from, keep in mind that some people may be sensitive to the proteins found in latex. Latex sensitivity is an issue that has prompted the glove industry to find alternative materials. Gloves are now made of materials such as vinyl, nitrile, and neoprene.

Perhaps the best place to begin when choosing appropriate hand protection is the ANSI/ISEA 105 Standard for Hand Protection Classification, developed by the American National Standards Institute and the International Safety Equipment Association. The standard addresses the classification and testing of hand protection for specific performance properties related to chemical and industrial applications.

For additional safety information, see the OSHA regulations regarding hand protection: **www.osha.gov/laws-regs/regulations/standardnumber/1910/1910.138**

Eyes on Safety

Nearly 500,000 eye injuries occur in the workplace every year in the United States alone. Experts say that 90 percent of those injuries could have been avoided if workers were more safety conscious and if they used the proper eye protection.

Breaking down these injuries, it adds up to more than 2,000 work-related eye injuries each day! Most injuries occurred while the workers were performing their regular jobs. Of those injuries, between 10 and 20 percent are disabling. This means the damage to one or both eyes were serious enough to result in temporary or even permanent vision loss.

OSHA reports that the majority of employees who injure their eyes either were not wearing any eye protection at the time of their accidents or were not wearing the right kind of protective eyewear for the particular job.

The top causes of eye injuries in the workplace are

- flying objects (bits of metal and glass),
- tools,
- dust and small particles,
- chemicals,
- harmful radiation, and
- a combination of these or other hazards.

Protective Eyewear Basics
- Always wear the proper eye safety gear. There are several types from which to choose, depending on the task you are performing:
  - Glasses
  - Goggles
  - Face shields
  - Welding helmets

- Follow all operating procedures correctly.

- Know where the first-aid and eye-cleaning stations are located and know how to use them properly.

- Always wear safety gloves and wash your hands after touching chemicals to prevent accidentally rubbing harmful substances into your eyes.

- Do not assume that wearing regular eyeglasses will protect your eyes. Regular eyeglasses are not designed to protect your eyes, and often they won't. Don't chance it.

- Make sure all protective eyewear fits properly and is not damaged. If it has been damaged, throw it away immediately.

Protective eyewear should be made of polycarbonate plastic. If you are working with liquids, your goggles should be splash-proof. Never rely on eyewear that is not designed for safety, such as reading glasses or sunglasses.

By taking a few safety precautions, you can greatly reduce your risk of eye injury. It takes only a few moments to think "eye safety" and put on safety goggles. It only takes a few seconds to use proper eye protection to protect yourself from a lifetime of problems.

For additional information, go to Prevent Blindness: **www.preventblindness.org**

Handling the Load: Forklift Safety

An often overlooked safety precaution is proper training for forklift operators. A forklift is the common name for a Powered Industrial Truck. Powered industrial trucks, commonly called forklifts or lift trucks, are used in many industries, primarily to move materials. They can also be used to raise, lower, or remove large objects or a number of smaller objects on pallets or in boxes, crates, or other containers. There are many types of powered industrial trucks. Each type presents different operating hazards. Even if your utility has instituted formal forklift operator training, a review of important practices will help you operate a forklift safely, both in a building and on the road.

- Only those individuals who are properly trained and qualified should operate a forklift.

- Always review the operator's manual prior to use of any forklift.

- Always inspect the vehicle at least once per shift. This includes checking the battery, brakes, controller, fuel system, horn, lights, lift system, steering mechanism, and tires. Don't operate any vehicle found to need repairs.

- Always use seat belts when operating forklifts.

- Look in the direction of travel and don't move the vehicle until you see your path is clear.

- Never allow passengers to ride in or on a forklift.

- Don't exceed the authorized safe speed.

- Don't pass trucks traveling in the same direction at intersections, blind spots, or other dangerous locations.

- Maintain at least three truck lengths' distance between you and vehicles in front of you.

- Slow down and sound the horn at cross aisles and other locations where vision is obstructed, use a flagger when possible.

- Carry the forks as low as possible.

- Cross over railroad tracks diagonally whenever possible.

- Don't load forklift trucks in excess of their rated capacity.

- Don't move a loaded vehicle until the load is secure.

- Carry a trailing load (i.e., drive in reverse with the load behind you) if the load would obstruct the view if carried in front of the vehicle.

- Ascend or descend a grade slowly. If the grade is in excess of 10 percent, drive a loaded truck with the load upgrade.

- Don't tilt a load forward with the load-engaging means elevated, except when picking up a load. Don't tilt an elevated load forward unless you are depositing it onto a storage area. When stacking or tiering, tilt a load backward only as much as necessary to stabilize the load.

- If you leave the vehicle and will be 25 ft (7.6 meters) or more away, leave the load-engaging means down, bring the mast to the vertical position, shut the power off, curb the vehicle if necessary, and set the brakes.

- If you leave the vehicle and are within 25 ft (7.6 meters) of the vehicle, lower the load-engaging means fully, neutralize the controls, and set the brakes.

Additional Notes

For more information, please refer to OSHA's regulations and resources related to Powered Industrial Trucks: **www.osha.gov/powered-industrial-trucks**

Reducing the Risk of Workplace Violence

Unhappy customers who harass and intimidate utility workers, either in a company location or in the field, pose a threat to those workers. OSHA singles out utility employees as being the most vulnerable to workplace violence. They are vulnerable because they deliver services, often work alone or in small groups, and may exchange money with the public. The most at-risk workers are the billing service staff, meter readers, and field staff who make house calls to investigate customer complaints or install services. Those responsible for shutting off water services are perhaps the most likely to encounter customer hostility.

Additionally, the department of Homeland Security has identified water and wastewater facilities as critical infrastructure, making them targets to threats and this can impact utility workers as well.

According to a survey conducted by Northwestern Mutual Life Insurance Company, 44 percent of workplace violence incidents are perpetrated by irate customers or clients. OSHA notes that workplace violence can occur anytime and anywhere.

However, once risk factors are assessed, occurrences can be prevented or minimized by knowing and using suitable precautions.

For utility workers, a potentially violent customer may catch a worker off guard. In this situation, a cool head and violence-prevention training come into play. A utility worker who encounters an angry customer at a company facility should never become defensive, confrontational, or patronizing. Instead, talk to the person in a calm, soft voice. This helps the person realize the volume of their own voice and may prompt that person to respond in kind.

- Listen closely to the complaint, smile pleasantly, and treat the customer with respect.

- Empathize by acknowledging how the person is feeling: "I understand why you are upset." "I know that this is difficult..."

- Ask open-ended questions: "What happened?" "What can we do to help you?"

By getting customers to talk instead of yell, you can break their train of thought and even diffuse their anger. No matter what, report the incident. Especially keep a record of volatile customers so other employees can be better prepared for future encounters.

In the field, all of the above suggestions apply. If the situation becomes uncomfortable, leave the premises, go to a safe place, and call for help. If the customer shows a weapon or physically threatens the utility worker, the incident needs to be immediately reported to the police as well as to utility management.

If a situation is potentially dangerous, such as shutting off service, OSHA recommends hiring an employee safety service or requesting police assistance. OSHA also recommends that employees who carry money should not work alone.

Other ways to increase staff safety include:

- Equipping field staff with cell phones, handheld alarms, or noise devices

- Requiring staff to set check-in times to keep a contact person informed of their location throughout the day

- Keeping utility vehicles in good working condition to avoid a breakdown in unsafe areas

- Providing drop safes to limit the amount of cash a bill collection employee carries.

- If a violent incident occurs, the employer should provide the affected employees with emotional support such as crisis intervention and counseling.

A workplace violence prevention program is only as effective as top management is willing to make it. But it is every employee's responsibility to be aware, act on warning signs, and learn how to deal with threats.

Utility workers should also have awareness about their personal security and following safeguards to prevent and mitigate threats as a critical infrastructure worker.

For additional information, go to the OSHA Safety and Health topics website: **www.osha.gov/SLTC/workplaceviolence**, or the Department of Homeland Security–Cyber and Infrastructure Security Agency (CISA) Personal Security Considerations Action Guide **www.cisa.gov/resources-tools/resources/ personal-security-considerations-action-guide**

Carbon Monoxide: A Silent Killer

- Water gushing from a 30-inch pipe near the University of California poured into Pauley Pavilion, and six people helping clean up the flooded arena were treated for carbon monoxide exposure from generator exhaust.

- Carbon monoxide leaking from a faulty flue pipe attached to a water heater killed the manager and sickened 27 others at a restaurant in New York.

- Downed power lines from ice storms in the Northeast and Midwest forced hundreds of thousands to spend the holidays without electricity, and carbon monoxide from gasoline-powered generators is blamed for eight deaths.

- A 77-year-old man was found dead in his home after leaving his car running in the garage.

These true stories are just a fraction of the deaths and illnesses reported every year from carbon monoxide (CO) poisoning. CO exposure can occur on the job as well as in homes and buildings that are inadequately ventilated and lack the proper detection devices. CO poisoning has affected people using gasoline-powered tools such as concrete cutting saws, high-pressure washers, floor buffers, welders, pumps, compressors, and generators. These incidents occur most often when these tools are operated indoors.

Carbon monoxide is an odorless, tasteless, colorless gas produced by the incomplete combustion of carbon-based fuels such as gasoline, natural gas, fuel oil, charcoal, or wood. Because of the potential for CO poisoning, small gasoline-powered engines and tools present a serious health hazard when operated indoors or in an enclosed space. CO can rapidly accumulate even in areas that appear to be well ventilated. Buildup can lead to dangerous or fatal concentrations within minutes. Opening doors and windows or operating fans does not guarantee safety.

Health Effects of Carbon Monoxide

Carbon monoxide interferes with the delivery of oxygen in the blood to the rest of the body. When you inhale high concentrations of CO, it can displace the oxygen in your bloodstream and cause one or more of the following symptoms:

- poor coordination,
- confusion and disorientation,
- fatigue,
- nausea,
- headache,
- dizziness,
- weakness,
- visual disturbances,
- changes in personality, and
- loss of consciousness.

If the concentration is high enough and the exposure is long enough, CO exposure can lead to death. About 1,000 people die each year from CO poisoning, according to the CDC.

Prevention Techniques

The CDC has the following recommendations to prevent CO poisoning in the workplace.

- Do not use or operate gasoline-powered engines or tools inside buildings or in partially enclosed areas.
- Learn to recognize the symptoms and signs of CO overexposure.
- Always place pumps, power units, and gasoline-powered compressors outdoors and away from air intakes so that engine exhaust is not drawn indoors where the work is being done.
- Consider using tools powered by electricity or compressed air if they are available and can be used safely.
- Use personal CO monitors where potential sources of CO exist. These monitors should be equipped with audible alarms to warn workers when CO concentrations are too high or when exceeding the NIOSH ceiling limit for CO of 200 parts per million.

- Conduct a workplace assessment to identify all potential sources of CO exposure.

- Educate workers about the sources and conditions that may result in CO poisoning and the symptoms and control of CO exposure.

- Monitor employee CO exposure to determine the extent of the hazard.

Additionally,

- always use the proper fuel in a combustion device, and

- don't leave a motor vehicle or gasoline-powered lawn mower running in enclosed spaces such as a garage or shed.

First Aid for CO Exposure

If you have any symptoms or notice that a coworker is impaired, immediately turn off the equipment and go outdoors or to a place with uncontaminated air.

- Call 911 or another local emergency number for medical attention or assistance if symptoms occur. Be sure to tell the first responder that you suspect carbon monoxide poisoning.

- Stay away from the work area until tools are deactivated and measured CO concentrations are below accepted guidelines and standards.

- Watch coworkers for the signs of CO toxicity.

- If you are affected by CO, do not drive a motor vehicle—get someone else to drive you to a health care facility.

For more information, go to the CDC for information on CO: **www.cdc.gov/niosh/topics/co-comp**

Night Work Safety

The majority of work at water utilities takes place during the day, but emergencies and the commitment to 24-hour, seven-days-a-week service means workers are also on the job at night. Working at night brings special challenges to maintaining a safe work environment.

Traffic Zones

For workers in traffic areas, the biggest challenge is finding a way to cope with the reduced visibility. At dawn and dusk, the sun is low in the sky and causes glare on a vehicle's windshield. Once the sun has set, the distance a motorist can see is restricted by headlight efficiency, and some drivers have poor night vision. As a result of these factors, crews working at night are three times more likely to be struck by a vehicle than their daytime counterparts.

Even when workers are wearing reflective safety vests, motorists aren't always able to determine that the object with the reflective tape is a person. When turned sideways, bending over, or standing motionless, workers are often mistaken for traffic cones or other safety markers. Motorists are less likely to slow down for a marker on the roadside than for a worker. Safety experts also tell us that working near the road is more dangerous at night because traffic is lighter, allowing motorists to travel faster through the work zone.

The condition of drivers at night also presents a hazard to workers. A higher percentage of drivers at night are subject to fatigue or to alcohol or drug impairment. Thirty-seven percent of drivers report having fallen asleep behind the wheel at some point, and an estimated 328,000 crashes each year in the United States involve drowsy drivers.

Here are some things you can do to make the work zone safer at night.

- Make sure your work clothing has an abundance of reflective material. The bright orange or yellow that motorists can see so well during the day does little good at night unless it is accompanied by reflective material on your vest or jacket, hard hat, and pants.

- Line up parked equipment so it serves as a boundary to protect work zones.

- Use floodlights to illuminate flagger stations, equipment crossings, and any other areas where crew members will be working. Floodlights can cause a disabling glare for drivers entering a work zone, so once the lights are set, a utility worker should drive through the area to observe their positioning and make adjustments as necessary.

- The Department of Transportation recommends that worksite lights should be clearly visible to drivers from a distance of 3,000 feet at night.

- Because of reduced visibility, crew members need to slow down and work more cautiously, especially when working around excavations. Shadows and dark areas inside trenches make the simple job of getting in and out of trenches more difficult. Footing near trench walls may appear to be more stable than it actually is.

- Crew members signaling and operating excavation equipment also need to take extra care in their job duties. The glare from traffic headlights and the fact that some excavation areas are partially hidden in shadows makes jobs more difficult.

Treatment Plants

Reduced visibility isn't just an issue at offsite work locations: because of dark areas and shadows created by floodlights, an area of the facility you are quite familiar with during daylight hours looks different at night. Outdoor filter beds, stairways and ramps, equipment storage areas, loading docks, and large water tanks are all areas that are more difficult to negotiate in the dark. Water storage tanks, for example, may be extra cold and have more moisture or ice on them at night, making footing or handholds more slippery and dangerous. Dew or ice may also exist on loading docks, stairways, and ramps, so slow down and take extra time and caution when walking across these areas.

When moving around the facility grounds at night, always carry a large flashlight with you to supplement whatever fixed lighting is available. It's a good idea to also carry a small backup flashlight in case the large light stops working during your rounds. Follow established paths using sidewalks, and when possible, walk in well-lit areas to avoid slip, trip or fall hazards. Mark trip hazards and low clearance areas with reflective paint or tape to prevent injuries.

Even though vehicular traffic is minimal on treatment plant grounds in the evening, you should still wear reflective clothing anytime you are outside the facility so coworkers and emergency personnel can see and identify you when they are on the facility grounds.

If you take the necessary precautions, your night-work duties can be performed without any problems. Don't get left in the dark; make the night shift safe and secure.

Avoid Harm From Laboratory Hazards

Water utility operators and personnel work in laboratory environments on a daily basis to complete daily process tests, compliance monitoring, and even optimization tasks. According to OSHA, these professionals are a part of more than 500,000 workers who are employed in laboratories in the United States.

Being in a laboratory can leave workers exposed to many hazards, including chemical, biological, and radioactive materials, as well as physical dangers.

When in a laboratory, keep yourself safe by remembering the following important steps.

- Think safety first.
- Know emergency responses.
- Know what you're working with.
- Use the smallest possible amount.
- Follow all safety procedures.
- If you don't know...ask!

Think Safety First

Engaging in horseplay or pranks can have devastating consequences in a laboratory. Always conduct yourself in a professional manner with constant self-awareness. Avoid cluttering workspaces, walkways, or exits with work materials to prevent safety hazards or a simple mix-up caused by disorganization.

Do not store food in laboratory fridges. Properly label chemical waste and in-use solutions with specific contents, and keep the label on the container. These practices are a part of good housekeeping.

Know Emergency Responses

Always alert others working in the laboratory immediately when a spill occurs or in an emergency situation. Do not clean up spills unless you are trained to do so.

It is important to promptly clean up spills, remembering to always wear your PPE. The supplies for cleaning up spills and any associated paperwork should be located in the immediate vicinity of the laboratory. Every lab should have eye-washing stations maintained properly in case of chemical ocular exposure.

Know What You're Working With

Always know the hazards for each material that is being used; if you are unsure check the Safety Data Sheet. When working with aerosols or volatile chemicals, use a fume hood.

Fume hood sashes should be kept closed as much as possible, and do not store chemicals in fume hoods.

Remember, it is better to be safe than sorry: treat every chemical as if it were hazardous.

Use the Smallest Possible Amount

Use the smallest amount of chemicals possible, but never return chemicals to the reagent bottles. Never mouth pipette, always use a bulb. Be aware of the various chemical exposure routes: dermal contact, inhalation, ingestion, ocular exposure, and injection.

Follow All Safety Procedures

Wear proper PPE and follow personal safety practices at all times when working in a laboratory. Lab coats, gloves, and safety glasses should be worn as appropriate. Shorts, sandals, or open-toed shoes should not be worn in the laboratory. Shoes should cover the entire foot. Glove selection must be suitable for the type of chemical substance in use to be resistant to permeation. No single glove material provides effective protection for all uses. It is best to secure any jewelry, loose clothing, or long hair before working to prevent any entanglement from occurring. Always wear proper eye protection when using chemicals. Never leave the laboratory space with your PPE or lab coat remaining on, when taking breaks.

If You Don't Know...Ask!

In all situations, ask if you are unsure of

- emergency procedures,
- laboratory rules,
- safety information,
- chemical locations,

- proper disposal of chemicals, and/or

- how to complete a task.

With any concern that you are not properly trained in or that you are confused about, always ask!

For more information, go to OHSA's guidance on the topic: **http://bit.ly/ OSHA3404**, or the National Academies Prudent Practices in the Laboratory: **http://nap.nationalacademies.org/read/12654/chapter/1**

Avoid Slips and Trips

Water utilities by their nature have many potential hazards that can cause slips, trips, and falls. These include slippery surfaces from water or liquid chemicals and tripping hazards, such as hoses, power cables, and irregular surfaces. US Department of Labor statistics show that slips, trips, and falls make up a majority of general industry accidents. Additionally, these types of incidents account for 15 percent of all accidental deaths and are the cause of 25 percent of all reported on-the-job injuries.

Source: Gino Santa Maria/Shutterstock.com

Reasons for Slips

Slips occur when there is too little friction or traction between feet (footwear) and the walking or working surface, resulting in loss of balance. Surfaces and situations that can cause slipping include the following:

- metal surfaces, such as ramps and gang planks;
- mounting and dismounting vehicles, ladders, and equipment;
- loose, irregular surfaces such as gravel;
- highly polished or waxed floors;
- transitioning from one surface to another, such as concrete to tile;
- sloped, uneven, or muddy walking surfaces;
- loose, unanchored rugs or mats;
- loose floorboards or shifting tiles;
- wet, muddy, or greasy shoes;
- dry product or wet spills; and
- natural hazards, such as ice, sand, and leaves and other plant debris.

Reasons for Trips

Trips happen when the moving foot of a person strikes an object, causing loss of balance. Situations and materials that contribute to trips include the following:

- uncovered hoses, cables, wires, or extension cords across aisles or walkways;
- clutter, obstacles in aisles, walkway, and work areas;
- open cabinet, file, or desk drawers and doors;
- changes in elevation or levels—as little as 1/4-in. difference can cause a trip;
- unmarked steps or ramps;
- rumpled or rolled-up carpets/mats or carpets with curled edges;
- irregularities in walking surfaces;
- thresholds or gaps;
- missing or uneven floor tiles and bricks;
- walking surface gratings improperly secured, damaged, or missing;
- uneven surfaces or objects protruding from walking surfaces;
- environmental conditions such as poor lighting, glare, shadows, excess noise, or temperature; and
- bulky PPE, including improper footwear.

While slips and trips are caused by irregularities in the pathway of a worker, inadequate awareness of those irregularities is a major contributor to most accidents. The human factor may be exacerbated by illness, poor vision, medications, or fatigue. Tripping and slipping can also be the result of carrying or moving cumbersome objects or too many objects at one time; walking while distracted by food, cellphones, or other devices; taking unapproved shortcuts; and rushing. All these factors can be controlled.

Institutional Control Measures

- Practice good housekeeping; maintain clear, tidy work areas free of clutter.

- Contain work processes to prevent discharge, splatter, or spillage of liquids, oils, particles, and dust onto walking surfaces.

- If obstacles can't be moved, mark them and reroute traffic around them.

- Secure all electrical and phone cords out of traffic areas; tape them to the floor or place them beneath a ramp.

- Keep work areas, aisles, stairwells, and pathways well lit.

- Mark/highlight step edges and transition areas (changes in elevations) with reflective tape and/or signage.

- Install slip-resistant floors in high-risk areas.

- Provide handrails along narrow or uneven walkways and stairs.

- Proper maintenance of guard rails, ladders, walkways/grating and stairways.

- Provide effective drainage on work platforms.

- Keep aisles and passageways clear of obstructions and in good repair.

- Clear outside areas of natural hazards such as leaves, loose gravel, and snow. Treat slippery surfaces such as ice with sand or salt.

- Ensure that mats and carpets have nonskid backing and that the edges aren't curling up.

- Install warning signs in areas prone to slipping, tripping, and falling hazards.

Personal Control Measures

- Follow safe routes—no shortcuts!

- Wear proper footwear with appropriate treads/traction for slip and chemical resistance.

- Don't wear sunglasses in low-light areas.

- Don't carry items that obstruct your view.

- Use guardrails and handrails.
- Slow down and pay attention to where you are walking!

For more information, see OSHA's recommendations on slips and falls: **www.osha.gov/ SLTC/etools/hospital/hazards/slips/slips.html**, or visit the National Safety Council website on fall prevention: **www.nsc.org/safety_home/HomeandRecreationalSafety/ Falls/Pages/Falls.aspx**, or visit OSHA's regulations on Walking-Working Surfaces: **www.ecfr.gov/current/title-29/subtitle-B/chapter-XVII/part-1910/subpart-D**

Be Prepared for an Emergency

Fire. Flood. Tornado. Hurricane. Storm surge. Earthquake. You never know when an emergency situation may force you to leave your home or workplace to deal with disaster situations. In the event of a major disaster, you and your family should realistically plan to be self-sufficient for at least seven days before outside resources are available. A little preparation now could save lives and prevent injuries in the future.

Escape routes. Every room in your house should have two designated escape routes. The whole family needs to know, understand, and practice the escape routes, especially children.

Evacuation plans. Massive evacuations caused by fire, hurricanes, and flooding are becoming more and more common. You may have only minutes to leave. So be ready to move!

- If you know there might be trouble soon, keep a full tank of gas in your car and only take one car per family to evacuate.

- Gather disaster supplies (see below) and have a battery-powered radio for official evacuation instructions. Don't forget the extra batteries!

- Before you leave, lock up your home and unplug everything except the freezer and refrigerator.

- Let others know where you're going, leave early to avoid being trapped, and follow recommended evacuation routes. Don't take shortcuts—they may be blocked!

Family communications. Your family may not be together when a disaster strikes, so plan how you will contact one another in emergency situations. Pick a friend or relative who lives out of state for family members to notify that they are safe.

Utility shutoff. Every adult needs to know how to shut off the utilities: natural gas, water, and electricity. Because different gas meter configurations have different shutoff procedures, contact your gas utility for guidance on preparation and response.

Food. Prepare in advance a week's worth of nonperishable food supplies for every family member. Try to avoid foods that will make you thirsty. Choose salt-free crackers, whole-grain cereals, and canned foods with high liquid content. Stock canned foods, dry mixes, and other staples that do not require refrigeration, cooking, water, or special preparation. You may already have many of these on hand. Be sure to include foods that meet special dietary needs. And do not forget a manual can opener!

Water. Water can become a precious resource after a disaster. Keep an emergency water sup-ply ample enough to meet the needs of the entire family for seven days or longer. Also plan on having enough water to meet your family's personal hygiene and sanitation needs.

Important documents. Store documents such as insurance policies, deeds, birth certificates, and property records in a bank safety deposit box away from home. Make copies for your disaster supply kit. Keep a small amount of cash or traveler's checks where you can quickly get to it.

Special needs. A family member with a disability or a special need may require additional assistance in an emergency. Find out what assistance is available in your community and be sure to inform the local office of emergency services and the fire department about your family's special needs.

Pets. If you must evacuate, don't leave your pets behind! They may not survive on their own, and you may not be able to find them when you return. Create a pet-survival kit that includes essential supplies such as food, water, and medications. For more information, contact the Humane Society of America.

Safety skills. Family members should know how to administer first aid and cardiopulmonary resuscitation (CPR). The American Red Cross frequently provides first-aid and CPR classes. Everyone should also know how to use a fire extinguisher; your home should have an ABC-type extinguisher.

Shelter. You may want to consider having sheltering supplies such as tarps, tents, and sleeping bags ready to go.

Emergency kit for work. This kit should be in one container and ready to grab and go in case you are evacuated from your workplace. Besides food and water in the kit, have comfortable walking shoes in case an evacuation requires you to walk long distances.

Emergency kit for your car. In case you are stranded, keep a kit of emergency supplies in your car. This kit should contain food, water, first-aid supplies, flares, jumper cables, and seasonal supplies.

Change stored food and water supplies in all your kits every six months and write the new date on all containers. You'll also need to rethink your supply needs every year and update your kit as your family needs change.

For more information, visit the Federal Emergency Management Agency website: **www.ready.gov**, the American Red Cross website: **www.redcross.org**, and your community's emergency service organizations.

Job Hazard Analysis:
Identify and Reduce Hazards

A job hazard analysis (JHA) is a safety evaluation process that focuses on the job task. Many companies, both large and small, have successfully used a JHA to identify potential dangers of specific tasks in order to reduce the risk of injury to workers.

It takes a little time to do a proper JHA, but it's time well spent. Be sure to involve employees in the process—they perform the work and often can easily identify the risks and know the best ways to work more safely.

- Selecting a job for a JHAStart by talking to your employees. Tell them what you are doing and why. Explain that you are studying the safety of the work tasks they perform and not their work performance.

- The frequency of performing a JHA can be beneficial for safety meetings to increase awareness, as part of an accident investigation to determine root cause failures, or part of a safety inspection to address safe work practices.

- Review your company's accident/injury/illness/near-miss history to determine which jobs pose the highest risk.

- Consider jobs or tasks that are most frequently completed or have the most complex aspects to complete that may require awareness for specific hazards and alternate various work groups such as operations or maintenance.

- Identify the OSHA standards that apply to your jobs and incorporate the OSHA requirements into the JHA.

- Evaluate jobs where you have identified violations of OSHA standards and/or violations of company safety procedures. List the jobs having the greatest potential to cause serious injuries or illness, even if there is no history of such problems.

- Make a note of the jobs in which a simple mistake could lead to severe injury.

- Evaluate jobs that are new or have been changed and those jobs that are so complex they require written instructions.

How to Conduct a JHA

Prepare to conduct a JHA by reviewing any applicable operational or maintenance manuals and documented procedures for the job, task, or equipment to be assessed.

Select a worker who is familiar with and normally performs the job or task to demonstrate the steps of the job or task. It may be a good idea to include the worker's supervisor to ensure the worker demonstrates proper procedure. Involving other parties who do not normally perform the job or task also increases awareness and allows for object questions to contribute to the JHA.

Watch each worker perform his or her job in a routine manner. List each step of each task in the order in which it takes place. Begin each step with a verb; for example, "Turn on the saw." Do not make the steps too broad or too detailed.

While you are making a record of the job, you may want to photograph or videotape each step for further analysis. Review the steps with all the workers who do the same job to make sure nothing's been left out.

Identify the hazards of each step and ask:

- What can go wrong?

- What are the consequences if something does go wrong?

- Consider specific categories of hazards to include physical hazards, chemical hazards, electrical hazards, or environmental hazards.

- How could an accident happen?

- Are there other contributing factors? The weather, seasonal workload, or new construction are examples.

- How likely is it that an accident will occur?

Review the List of Hazards With Employees

Your employees can provide a tremendous amount of information. Take the time to talk to them—and be sure to listen. Asking for their honest input will engage them in the process and lead to a higher level of safety awareness. You will likely hear several practical ways they believe the job hazards and job processes affecting them can be eliminated or at least reduced.

Eliminate or Reduce the Safety Hazards

You've evaluated the findings in your analysis and concluded there's a safer way to do the job. Now your work begins.

- Always follow the hierarchy of controls when considering safeguards to address specific hazards: Elimination, Substitution, Engineering Controls, Administrative Controls, and Personal Protective Equipment (PPE).

- First, make any changes to the equipment, tools, or engineering controls to eliminate a hazard. Such changes might include adding machine guards, improving lighting, or having better ventilation.

- Change the work processes.

- Change the administrative controls or make changes to how the task is done if engineering controls aren't possible. Perhaps you could rotate jobs, change the steps in the process, or provide additional training.

- When engineering and administrative controls aren't possible or don't adequately protect the workers, make additions and changes to the required personal protective equipment.

Implement Your JHA Changes

To complete the JHA, you'll need to correct all unsafe conditions and processes. The resulting changes may require additional training for your employees. Make sure they understand the changes and the reasons behind those changes.

Periodically review the JHA. A JHA is a snapshot in time. You may find hazards you missed before, or conditions and processes may change. Update and review the document when the task or process changes or when injuries or a close call occurs when performing the recommended task.

You'll find your JHA to be a valuable tool. Not only will it help to reduce worker injuries, but it's a document you can use for training purposes or for standard operating procedure development. It can also serve as a reference tool in the event of an accident investigation.

For additional information, read the OSHA booklet *Job Hazard Analysis*: **www.osha. gov/Publications/osha3071.pdf**, or see AWWA M3, *Safety Management for Utilities*.

Avoid Arc Flash

Arc flashes or blasts pose a significant danger when working on or around electricity. Arc hazards in a water utility are most likely to come from switchboards, panel boards, and motor and industrial control centers. Workers at risk are those examining, servicing, or providing maintenance on these components.

Temperatures during an arc flash can reach as high as 35,000 degrees—nearly four times the temperature of the surface of the sun. Two thousand people each year are admitted to burn centers with severe arc flash injuries.

Arc flashes can injure or kill workers at distances of 15–20 ft. An arc flash can burn the skin directly and ignite a worker's clothing. Shrapnel, molten metal droplets, and particles are all dangerous elements of an arc flash or blast. These incidents can also result in hearing and respiratory damage, as well as eye and face injuries.

The threat goes beyond just the person working on the electrical piece of equipment; because arc flashes are so large and powerful, anyone in the immediate area is at risk.

Source: Reggie Lavoie/Shutterstock.com

157

How Arc Flashes Happen

An arc flash or burst occurs primarily while someone is working on an energized circuit. A flash can occur spontaneously or from bridging electrical contacts with a conducting object. This can happen if a worker drops a tool or accidently makes contact with the equipment. Excessive corrosion or dust buildup on the contact points can also spark an arc flash. An arc flash or blast can also occur simply because of an electrical equipment malfunction or failure.

Specific OSHA and National Fire Protection Association (NFPA) regulations and recommendations address arc flash safety on the work site. For example, labeling of electrical equipment advising workers when "a dangerous condition associated with the possible release of energy caused by an electric arc" exists. OSHA mandates that only qualified persons are permitted to work on electrical conductors and circuits.

Avoiding Arc Flash

Have a written plan and permit system for conducting any work on or near energized equipment of more than 50 volts. The permit should list required conditions and work practices specific to the location of the work, the circuit and equipment involved, the hazard analysis, required PPE and tools, safe work practices, access control, and boundaries for approach by other workers.

Conduct electrical safety training. All impacted workers who may be exposed to arc flash hazards in the workplace need to have awareness of the hazards they are subject to and what safeguards must be followed to prevent injury.

Conduct a flash hazard analysis. Flash arc hazard boundaries and limits of approach are based on the voltage and is calculated using various formulas. It's important to establish and ensure an electrically safe work area, maintained throughout the work period. Properly test for voltage and grounding power conductors.

De-energize electrical equipment. Begin by thoroughly identifying all power sources. Then disconnect or interrupt that service when possible with a visual verification of the open circuit.

Follow proper lockout/tagout procedures. Visually verify the disconnect has opened the circuit, apply lockout tagout devices, test for the absence of voltage, and use ground phase conductors to counteract stored energy and induced voltage.

Wear appropriate PPE. Depending on the voltage present, arc flash safety guidelines may require safety glasses, hearing protection, flame-resistant clothing, a full flash suit, a face shield, a switching coat and hood, shoes and gloves. Protective clothing is "arc rated" depending on the anticipated hazard. All PPE must be inspected and tested in accordance with manufacturer recommendations prior to use.

Use the proper tools. Use only double insulated tools. A high-visibility yellow layer provides insulation for the tool, and an outer high-visibility orange layer protects the lower layer. If the yellow underneath layer can be seen, the tool should be removed from service. These tools generally have a maximum safety rating of up to 1,000 volts to protect you from accidental contact but are not designed to be used on energized circuits. Examine your tools before each use, keep them clean and dry, and have a qualified person recertify them periodically.

Additionally, when working with potentially energized equipment,

- position your body to the side and away as much as possible during switching,

- avoid touching switchgear and metallic surfaces, and

- use metal-clad and arc-resistant switchgear and current-limiting power circuit breakers and reactors.

For more information, see NFPA 70E *Standard for Electrical Safety in the Workplace*: **www.nfpa.org/codes-and-standards/nfpa-70e-standard-development/70e**

Additional AWWA Safety Products

To order any of these products or for more information, call our customer service line at 1-800-926-7337 or visit our online store at **http://store.awwa.org**.

Manuals

M3 Safety Management for Utilities (order #30003-8E)

This manual M3 focuses on the development of comprehensive health, safety, and environment (HSE) programs. M3 can be used as a general guidance for personnel who have the responsibility for developing, implementing, and monitoring their utility's HSE program. Supervisors, who have safety obligations to their employees, will also find M3 useful, particularly with the numerous descriptions of and common controls for typical hazards faced by staff on a routine basis.

M19 Emergency Planning for Water and Wastewaster Utilities (order #30019-5E)

This manual outlines principles, practices, and guidelines for water utility emergency planning such as natural disasters, accidents, or intentional malevolent acts.

Standards

ANSI/AWWA G430 *Security Practices for Operation and Management* (order #47430-2020)

ANSI/AWWA G440 *Emergency Preparedness Practices* (order #47440-2017)

ANSI/AWWA J100 *Risk Analysis and Management for Critical Asset Protection* (RAMCAP®) *Standard for Risk and Resilience Management of Water and Wastewater Systems* (order #40100-2021)

This standard describes the application of RAMCAP, a seven-step process for identifying, analyzing, and managing risks associated with malevolent attacks against and naturally occurring hazards affecting our nation's critical infrastructure.

Video Streaming

Safety First **Channel**

Safety First on video streaming is the solution to effective and professional safety training. It's easy to train anywhere, anytime. These high-resolution videos provide comprehensive and engaging safety training on a large selection of safety topics, all specifically focused for water utilities. **awwa.org/training-videos**